Martin Loebl

Discrete Mathematics in Statistical Physics

Advanced Lectures in Mathematics

Geometric Graphs and Arrangements
Stefan Felsner

Ruled Varieties
Gerd Fischer, Jens Piontkowski

Dirac-Operatoren in der Riemannschen Geometrie
Thomas Friedrich

Discrete Mathematics in Statistical Physics
Martin Loebl

The Steiner Tree Problem
Hans Jürgen Prömel, Angelika Steger

Algebraic Geometry (Part 1 and Part 2)
Ulrich Görtz, Torsten Wedhorn
(in preparation)

www.viewegteubner.de

Martin Loebl

Discrete Mathematics in Statistical Physics

Introductory Lectures

**VIEWEG+
TEUBNER**

Bibliographic information published by the Deutsche Nationalbibliothek
The Deutsche Nationalbibliothek lists this publication in the Deutsche Nationalbibliografie;
detailed bibliographic data are available in the Internet at http://dnb.d-nb.de.

Dr. Martin Loebl
Charles University
Department of Applied Mathematics
Institut of Theoretical Computer Science
Malostranske 25
CZ-118 00 Praha
Tschechische Republik

E-Mail: loebl@kam.mff.cuni.cz

1st Edition 2010

All rights reserved
© Vieweg+Teubner | GWV Fachverlage GmbH, Wiesbaden 2010

Editorial Office: Ulrike Schmickler-Hirzebruch | Nastassja Vanselow

Vieweg+Teubner is part of the specialist publishing group Springer Science+Business Media.
www.viewegteubner.de

Cover design: KünkelLopka Medienentwicklung, Heidelberg
Printing company: STRAUSS GMBH, Mörlenbach
Printed on acid-free paper
Printed in Germany

ISBN 978-3-528-03219-7

dedicated to Zbyněk and Jaroslava, my parents

Preface

The purpose of these lecture notes is to briefly describe some of the basic concepts interlacing discrete mathematics, statistical physics and knot theory. I tried to emphasize a 'combinatorial common sense' as the main method. No attempt of completeness was made. The book should be accessible to the students of both mathematics and physics. I profited from previous books and expositions on discrete mathematics, statistical physics, knot theory and others, namely [B1], [BRJ], [BB], [J1], [KG], [LL], [MN], [MJ], [MT], [S0], [S3], [SM], [WFY], [WD], [KSV]. Most of the material contained in the book is introductory and appears without a reference to the original source. This book has been an idea of my editor Martin Aigner. I would like to thank to him for his support and help. Many other colleagues helped me with the book. Mihyun Kang, Jirka Matoušek, Iain Moffatt, Jarik Nešetřil, Dominic Welsh and Christian Krattenthaler read earlier versions, and without their extensive comments the book would probably not exist. I had enlightening discussions on several topics discussed in the book, in particular with Martin Klazar, Roman Kotecký, Ondřej Pangrác, Gregor Masbaum, Xavier Viennot and Uli Wagner. Marcos Kiwi saved the whole project by gently teaching me how to draw pictures amd Winfried Hochstaettler drew one; I am sure you will be able to detect it. Large part of the book was written during my visit, in the whole year 2006, at the School of Mathematics and the Centro Modelamiento Matematico, Universidad de Chile. I want to thank my colleagues there for wonderful hospitality, and gratefully acknowledge the support of CONICYT via project Anillo en Redes, ACT-08. But of course, the seminal ingredient in the process of making the book was the creative environment of my home department of applied mathematics and the institute of theoretical computer science at the Charles University, Prague.

Some theorems and observations in the book appear without a proof. Usually a pointer is given to a book (preferentially) or to a paper where a proof can be found. If no pointer is given, then I believe (possibly mistakenly) that it should be possible to prove the statement in an elementary and not very complicated way. The reader is encouraged to write down such proofs as exercises. The first five chapters concentrate on the introductory discrete mathematics. Chapters six and seven are devoted to the partition functions, and chapter eight is an introduction to the theory of knots. The last chapter describes two combinatorial technics which solve the 2D Ising and dimer problems.

Prague, September 2009
Martin Loebl

Contents

Chapter 1

Basic concepts

In this introductory chapter we first present some very basic mathematical formalism. Then we introduce algorithms and complexity. The chapter ends with basic tools of discrete calculations.

1.1   Sets, functions, structures

We will use symbols $\mathbb{N}, \mathbb{Z}, \mathbb{P}, \mathbb{Q}, \mathbb{R}$, and $\mathbb{C}$ to denote the sets of the natural, integer, positive integer, rational, real, and complex numbers. If not specified otherwise then i, j, k, m, n are non-negative integers. We sometimes denote by $[n]$ the set $\{1, \ldots, n\}$. We denote by $|X|$ the cardinality of a set X. A function f from a set X to a set Y is called *one-to-one* or *injective* if for $x, y \in X$ $x \neq y$ implies $f(x) \neq f(y)$, it is *surjective* or *onto* if for each $y \in Y$ there is $x \in X$ such that $f(x) = y$, and it is a *bijection* if it is both one-to-one and onto. If X, Y are finite then a bijection is also called a *permutation*. Let $|X| = n$ and $|Y| = m$. There are $n! = n(n-1)(n-2)\cdots 1$ permutations of X; $n!$ is the *factorial function*. The number of all functions from X to Y is m^n and the number of one-to-one functions from X to Y is $m(m-1)\cdots(m-n+1)$. The number of surjective functions does not have such a nice formula, it may be written with the help of the principle of inclusion and exclusion as

$$\sum_{i=0}^{m} (-1)^i \binom{m}{i} (m-i)^n. \tag{1.1}$$

The *Kronecker delta function* is defined by $\delta(x, y) = 1$ if $x = y$, and zero otherwise. If X is a set, then we denote by 2^X the set of all the subsets of X; $|2^X| = 2^n$. We further denote by $\binom{X}{k}$ the set of all subsets of X of cardinality k. We have

$$\left| \binom{X}{k} \right| = \binom{n}{k} = \frac{n!}{k!(n-k)!}.$$

The *binomial theorem* says that

$$(x + y)^n = \sum_{k=0}^{n} \binom{n}{k} x^k y^{n-k}.$$

The symbol $\binom{n}{k}$ is called the *binomial coefficient*. The *multinomial coefficient* is defined by

$$\binom{n}{k_1, \cdots, k_m} = \frac{n!}{k_1! \cdots k_m!}$$

and the *multinomial theorem* says that

$$(x_1 + \cdots + x_m)^n = \sum_{k_1 + \cdots + k_m = n} \binom{n}{k_1, \cdots, k_m} x_1^{k_1} \cdots x_m^{k_m}.$$

A very good estimate of the factorial function $n!$ is given by Stirling's formula which approximates $n!$ by $(2\pi n)^{1/2}(\frac{n}{e})^n$.

If $Y \subset X$, then the *incidence vector* of Y will be denoted by $i(Y)$; $i(Y)$ is the $0, 1$ vector indexed by the elements of X, where $[i(Y)]_z = 1$ if and only if $z \in Y$. We will sometimes not distinguish between a set and its incidence vector.

An *ordered pair* is usually denoted by (x, y) where x is the first element of the pair. A *binary relation* on X is any subset of $X \times X = \{(x, x'); x \in X, x' \in X\}$. Any function $X \to X$ is a binary relation on X. A *partially ordered set*, or *poset* for short, is a pair $(X, \preceq)$, where X is a set and $\preceq$ is a binary relation on X that is reflexive ($x \preceq x$), transitive ($x \preceq y$ and $y \preceq z$ imply $x \preceq z$), and (weakly) antisymmetric ($x \preceq y$ and $y \preceq x$ imply $x = y$). The binary relation $\preceq$ is itself called a *partial ordering*. A partial ordering where any pair of elements is comparable is called a *linear ordering*. An important example of a linear ordering is the *lexicographic ordering*. Let $a = (a_1, \ldots, a_n)$ and $b = (b_1, \ldots, b_m)$ be two strings of integers. We say that a is lexicographically smaller than b if a is an initial segment of b or $a_j < b_j$ for the smallest index j such that $a_j \neq b_j$. Let $(X, \preceq)$ be a poset and $Y \subset X$. We say that Y is a *chain* if the induced ordering $(Y, \preceq)$ is linear.

The symbol $\mathbb{F}$ will denote a field; $\mathbb{F}$ will usually be equal to $\mathbb{R}$ or $\mathbb{C}$, or to the finite 2-element field $GF(2) = (\{0, 1\}, +, \times)$ with addition and multiplication modulo 2. The symbol $\mathbb{F}^d$ denotes the vector space of dimension d over $\mathbb{F}$. The elements of $\mathbb{F}^d$ are called vectors; for $x \in \mathbb{F}^d$ we write $x = (x_1, \cdots, x_d)$. We will understand vectors as both row and column vectors. The scalar product of two vectors x, y is $xy = x_1 y_1 + \cdots + x_d y_d$. A set $\{x^1, \cdots, x^k\}$ of vectors of $\mathbb{F}^d$ is *linearly independent* if, whenever $\sum_{i=1}^{k} a_i x^i = 0$ and each $a_i \in \mathbb{F}$, then $a_1 = a_2 = \cdots = a_k = 0$. The *dimension* $\dim(\{x^1, \cdots, x^k\})$ of a set of vectors $\{x^1, \cdots, x^k\}$ is the maximum number of linearly independent elements in $\{x^1, \cdots, x^k\}$. A *subspace* of a vector space $\mathbb{F}^d$ is any subset of $\mathbb{F}^d$ which is closed under addition, and multiplication by an element of $\mathbb{F}$. Two subspaces X, Y are *isomorphic* if there is a bijection $f : X \to Y$ such that for each $a, b \in X$ and $c \in \mathbb{F}$, $f(a + b) = f(a) + f(b)$ and $f(ca) = cf(a)$. The *orthogonal complement* of a subspace $X \subset \mathbb{F}^d$ is the subspace $\{y \in \mathbb{F}^d; xy = 0 \text{ for each } x \in X\}$.

Let $A = (a_{ij})$ be a matrix of n rows and m columns, with entries from field $\mathbb{F}$. We say that A is an $n \times m$ matrix. If $n = m$ then A is a *square matrix*. The *determinant* of a square $n \times n$ matrix A is defined by $\det(A) = \sum_\pi (-1)^{\text{sign}(\pi)} \prod_{i=1}^n a_{\pi(i)i}$, where the sum is over all permutations π of $1, \cdots, n$ and $\text{sign}(\pi) = |\{i < j; \pi(i) > \pi(j)\}|$. The determinant characterizes linearly independent vectors. A set of n vectors of length n is linearly independent if and only if $\det(A) \neq 0$, where A is the matrix whose set of columns (or rows) is formed by the vectors. The determinant of a matrix may be calculated efficiently by the *Gaussian elimination*. The *permanent* of matrix A is defined analogously as the determinant, but the $(-1)^{\text{sign}(\pi)}$ term is omited from each summand. Hence, $\text{Per}(A) = \sum_\pi \prod_{i=1}^n a_{\pi(i)i}$. There is no efficient algorithm to calculate the permanent. The *identity matrix* is the square matrix A such that $a_{ii} = 1$ and $a_{ij} = 0$ for $i \neq j$. The *trace* of a square matrix A, denoted by $\text{tr}(A)$, is defined by $\text{tr}(A) = \sum_i a_{ii}$.

The symbol $\mathbb{R}^d$ also denotes the Euclidean space of dimension d. A *curve* in $\mathbb{R}^d$ is the image of a continuous function $f : [0, 1] \to \mathbb{R}^d$. A curve is *simple* if it is one-to-one, and it *connects* its endpoints $f(0), f(1)$. A curve is *closed* if $f(0) = f(1)$. The *Euclidean norm* of $x \in \mathbb{R}^d$ is $||x|| = (xx)^{1/2}$. A set $\{x^0, x^1, \cdots, x^k\}$ of vectors of $\mathbb{R}^d$ is *affinely independent* if, whenever $\sum_{i=0}^k a_i x^i = 0$, $\sum_{i=0}^k a_i = 0$ and each $a_i \in \mathbb{R}$, then $a_0 = a_1 = \cdots = a_k = 0$. For two points x_0, x_1 affine independence means $x^0 \neq x^1$; for three points it means that x^0, x^1, x^2 do not lie on a common line; for four points it means that they do not lie on a common plane; and so on. The *rank* of a set of points of $\mathbb{R}^d$, denoted by $\text{rank}(\{x^0, \cdots, x^k\})$, is the maximum number of affinely independent elements in $\{x^0, \cdots, x^k\}$.

There is a simple relation between linear and affine independence: $x^0, \cdots, x^k$ are affinely independent if and only if $x^1 - x^0, \cdots, x^k - x^0$ are linearly independent. This happens if and only if the $(d+1)$-dimensional vectors $(1, x^0), \ldots, (1, x^k)$ are linearly independent. An *affine subspace* is any subset $A \subset \mathbb{R}^d$ which contains, for each pair of its elements x, y, the line through x, y. A *hyperplane* in $\mathbb{R}^d$ is a $(d - 1)$-dimensional affine subspace, i.e., a set of the form $\{x \in \mathbb{R}^d : ax = b\}$ for some nonzero $a \in \mathbb{R}^d$ and $b \in \mathbb{R}$. A (closed) *half-space* has the form $\{x \in \mathbb{R}^d : ax \leq b\}$ for some nonzero $a \in \mathbb{R}^d$ and $b \in \mathbb{R}$.

A set $C \subset \mathbb{R}^d$ is *convex* if for every $x, y \in C$, the segment $\{ax + (1-a)y : 0 \leq a \leq 1\}$ between x and y is contained in C. The *convex hull* of a set $X \subset \mathbb{R}^d$ is the intersection of all convex sets containing X, and it is denoted by $\text{conv}(X)$. Each $x \in \text{conv}(X)$ may be written as a *convex combination* of elements of X: there are $x^1, \cdots, x^k \in X$ and real numbers $a_1, \cdots, a_k \geq 0$ such that $\sum_{i=1}^k a_i = 1$ and $x = \sum_{i=1}^k a_i x^i$.

A *convex polytope* is the convex hull of a finite subset of $\mathbb{R}^d$. Each convex polytope can be expressed as the intersection of finitely many half-spaces. Conversely, by the Minkowski-Weyl theorem, if an intersection of finitely many half-spaces is bounded, then it is a convex polytope. A *face* of a convex polytope P is P itself or a non-empty intersection of P with a hyperplane that does not dissect P (i.e., not both of the open half-spaces defined by the hyperplane intersect P in a non-empty set).

1.2   Algorithms and Complexity

Algorithmic considerations are important for many concepts of both discrete mathematics and statistical physics. We make only basic algorithmic remarks in this book, and therefore the following exposition on algorithms and complexity is very brief.

Informally, an *algorithm* is a set of instructions to be carried out mechanically. Applying an algorithm to its *input* we get some *output*, provided that the sequence of the instructions prescribed by applying the algorithm terminates. The application of an algorithm is often called a *computation*. Usually inputs and outputs are strings (words, finite sequences) from a finite alphabet; a basic example are binary words, i.e., finite sequences of $0, 1$. The notion of an algorithm is usually formalized by the definition of a Turing machine.

A *Turing machine* consists of the following components:

- a finite set S called the *alphabet*,

- an element $b \in S$ called the *blank symbol*,

- a subset $A \subset S$ called the *external alphabet*; we assume $b \notin A$,

- a finite set Q whose elements are called *states* of the Turing machine,

- an initial state $s \in Q$,

- a *transition function*, i.e., a function

$$t : Q \times S \to Q \times S \times \{-1, 0, 1\}.$$

A Turing machine has a *tape* that is divided into cells. Each cell carries one symbol from S. We assume that the tape is infinite, thus the content of the tape is an infinite sequence $s = s_0, s_1, \cdots$ of elements of S.

A Turing machine also has a read-write *head* that moves along the tape and changes symbols. If the head is in position p along the tape, it can read symbol s_p and write another symbol in its place.

The behaviour of a Turing machine is determined by a control device. At each step of the computation, this device is in some state $q \in Q$. The state q and the symbol s_p under the head determine the action performed by the Turing machine: the value of the transition function, $t(q, s_p) = (q', s', p')$, contains the new state q', the new symbol s' to be written in the place of s_p, and the shift $p' \in \{-1, 0, 1\}$ of the position of the head. If the head bumps into the left boundary of the tape (that happens when $p + p' < 0$), then the computation *stops*.

Next we describe the input given to the Turing machine, and how the output is obtained. Let A^* denote the set of all the strings (finite sequences) of elements of A. Inputs and outputs to the Turing machine with the external alphabet A are strings from A^*. An input string I is written on the tape and followed by the blank symbol b. Initially, the head is at the beginning (left end) of I. If the Turing machine stops (by bumping into the left boundary of the tape), we read the tape from left to right starting from the left end until we reach some symbol that does not belong to A. The initial segment of the tape until that symbol will be the output of the Turing machine.

Every Turing machine *computes* a function from a subset of A^* to A^*. There are functions that are *not computable*. A Turing machine is obviously an algorithm in the informal sense. The converse assertion is called the

Church-Turing thesis: Any algorithm can be realised by a Turing machine. Note that the Church-Turing thesis is not a mathematical theorem, but rather a statement about our understanding of the informal notion of algorithm.

Complexity classes. The computability of a function does not guarantee that we can compute it in practice since an algorithm may require too much time. The idea of an effective algorithm is usually formalized by the notion of *polynomial algorithms*. We say that a function T on the positive integers is of *polynomial growth* if $T(n) \leq cn^d$ for all n and some constants c, d. We say that a function f defined on the binary strings of $\{0,1\}^*$ is *computable in polynomial time* if there exists a Turing machine that computes f in time $T(n)$ of polynomial growth, where n is the length of the input. Such a Turing machine is called a *polynomial algorithm*. Polynomial time *encoding* plays a crucial role. For instance, if the input is an integer N in the *unary* representation then the input size is $|N|$ but if the representation is binary, the input size is only $\log(|N|)$. The class of all functions computable in polynomial time is denoted by P. We should remark here that computability in polynomial time does not guarantee practical computability either, but it is a good indication for it.

A special class of algorithmic problems are the *decision problems*. In a decision problem, we want the answer to be *yes* or *no*. This clearly may be modeled as a function from a subset of A^* to $\{0,1\}$ where 0 encodes *no* and 1 encodes *yes*. It is customary to call such functions *predicates*. One can think about predicates as about properties: the predicate indicates for each string whether it has the property (yes) or does not have the property (not). Hence the algorithmic problem to compute a predicate may be formulated as the algorithmic problem to test the corresponding property.

Another basic complexity class, the class NP, is usually defined only for the predicates. We say that a predicate $R(x,y)$, where x and y are binary strings, is *polynomially decidable* if there is a Turing machine that computes it in time of polynomial growth (the size of the input is $|x| + |y|$).

The class NP is the class of all predicates f for which there is a polynomial growth function $T(n)$ and a polynomially decidable predicate R of two variables so that $f(x) = 1$ if and only if there is y such that $|y| < T(|x|)$ and $R(x,y) = 1$. Informally, NP is the class of the predicates (i.e., properties), for which there is a certificate (coded by y) that can be checked in polynomial time. Most of the properties discussed in this book belong to NP.

Clearly $P \subset NP$. Over the past 30 years intensive research has been directed towards proving that the inclusion is strict. The question whether $P \neq NP$ is today one of the fundamental problems of both mathematics and computer science.

Reducibility. When can we say that one problem is algorithmically at least as hard as another problem? We model the efficiency by the polynomial time complexity, and so the answer is naturally given by the following notion of *polynomial reducibility*: we say that a predicate f_1 is *polynomially reducible* to a predicate f_2 if there exists a function $g \in P$ so that $f_1(x) = f_2(g(x))$, for each input string x.

A predicate $f \in NP$ is called NP-*complete* if any predicate in NP is polynomially reducible to it. The predicates that are NP-complete are the most difficult predicates of NP: if some NP-complete predicate is in P then $P = NP$. It is customary to speak about NP-*complete problems* rather than NP-complete predicates. The existence of an efficient algorithm to solve an NP-complete problem is considered to be very unlikely.

A seminal result in algorithmic complexity is that NP-complete predicates (problems) exist. This was proved independently by Cook and Levin. Many natural NP-complete problems are known, see [GJ].

1.3   Generating functions

A useful way of counting is provided by generating functions. If f is a function from the non-negative integers, we can consider its (ordinary) *generating function* $\sum_{n \geq 0} f(n)x^n$ and its *exponential generating function* $\sum_{n \geq 0} f(n)x^n/n!$.

The generating functions are *formal power series*, since we are not concerned with letting x take particular values, and we ignore questions of convergence. This formalism is convenient since we can perform various operations on the formal power series, for instance

$$\left(\sum_{n \geq 0} a_n x^n \right) + \left(\sum_{n \geq 0} b_n x^n \right) = \sum_{n \geq 0} (a_n + b_n)x^n,$$

$$\left(\sum_{n \geq 0} a_n x^n/n! \right) + \left(\sum_{n \geq 0} b_n x^n/n! \right) = \sum_{n \geq 0} (a_n + b_n)x^n/n!$$

and

$$\left(\sum_{n \geq 0} a_n x^n \right) \left(\sum_{n \geq 0} b_n x^n \right) = \sum_{n \geq 0} c_n x^n$$

$$\left(\sum_{n \geq 0} a_n x^n/n! \right) \left(\sum_{n \geq 0} b_n x^n/n! \right) = \sum_{n \geq 0} d_n x^n/n!$$

where $c_n = \sum_{i=0}^n a_i b_{n-i}$ and $d_n = \sum_{i=0}^n \binom{n}{i} a_i b_{n-i}$.

These operations coincide with the addition and multiplication of functions when the power series converge for some values of x. Let us denote by $\mathbb{C}[[x]]$ the set of all formal power series $\sum_{n\geq 0} a_n x^n$ with complex coefficients. Addition and multiplication in $\mathbb{C}[[x]]$ are clearly commutative, associative and distributive, thus $\mathbb{C}[[x]]$ forms a commutative ring where 1 is the unity. Formal power series with the coefficients in a non-commutative ring (like the square matrices of the same size) are also extensively considered; they form a non-commutative ring with unity.

If $F(x)$ and $G(x)$ are elements of $\mathbb{C}[[x]]$ satisfying $F(x)G(x) = 1$ then we write $G(x) = F(x)^{-1}$. It is easy to see that $F(x)^{-1}$ exists if and only if $a_0 = F(0) \neq 0$. If $F(x)^{-1}$ exists then it is uniquely determined. We have $((F(x)^{-1})^{-1} = F(x)$.

Example 1.3.1. Let $a \neq 0$ and $(\sum_{n\geq 0} a^n x^n)(1 - ax) = \sum_{n\geq 0} c_n x^n$, where a is a non-zero complex number. Then from the definition of multiplication we get $c_0 = 1$ and $c_n = 0$ for $n > 0$. Hence we may write

$$\sum_{n\geq 0} a^n x^n = (1 - ax)^{-1}.$$

The identity may be derived in the same way in every ring of formal power series over a (not necessarily commutative) ring with unity. Hence, for instance, for square complex matrices it can be written as

$$\sum_{n\geq 0} A^n x^n = (I - Ax)^{-1}.$$

This is of course just the formula for summing a geometric series. Informally speaking, if we have an identity involving power series that is valid when the power series are regarded as functions (when the variables are sufficiently small complex numbers), then the identity remains valid when regarded as an identity among formal power series. Formal power series may naturally have more than one variable.

1.4   Principle of inclusion and exclusion

Let us start with the introduction of a paper of Whitney, which appeared in Annals of Mathematics in August 1932:

"Suppose we have a finite set of objects (for instance books on a table), each of which either has or has not a certain given property A (say of being red). Let n be the total number of objects, $n(A)$ the number with the property A, and $n(\bar{A})$ the number without the property A. Then obviously $n(\bar{A}) = n - n(A)$. Similarly, if $n(AB)$ denotes the number with both properties A and B, and $n(\bar{A}\bar{B})$ the number with neither property, then $n(\bar{A}\bar{B}) = n - n(A) - n(B) + n(AB)$, which is easily seen to be true. The extension of these formulas to the general case where any number of properties is considered is quite simple, and is well known to logicians. It should be better known to mathematicians also; we give in this paper several applications which show its usefulness."

It is known today, under the name *principle of inclusion and exclusion* (PIE).

Theorem 1.4.1. *Suppose $A_1, \ldots, A_n$ are finite sets, and $A_J = \bigcap_{i \in J} A_i$. Then*

$$\left| \bigcup_{i=1}^{n} A_i \right| = \sum_{k=1}^{n} (-1)^{k-1} \sum_{J \in \binom{[n]}{k}} |A_J|.$$

Proof. We proceed by induction on n. The case $n = 2$ is clear. In the induction step,

$$\left| \bigcup_{i=1}^{n} A_i \right| = \left| \bigcup_{i=1}^{n-1} A_i \cup A_n \right| =$$

$$|A_n| + \sum_{k=1}^{n-1} (-1)^{k-1} \Sigma_{J \in \binom{[n-1]}{k}} |A_J| - \left| \bigcup_{i=1}^{n-1} (A_i \cap A_n) \right| =$$

$$\sum_{k=1}^{n} (-1)^{k-1} \Sigma_{J \in \binom{[n]}{k}} |A_J|.$$

□

Let $(X, \leq)$ be a finite partially ordered set (poset). For example, the set of all subsets of a finite set S equipped with the relation '$\subset$' forms a poset called the *Boolean lattice*. Let $\mathbb{F}$ be a field and let us denote by $\mathcal{F}(X)$ the collection of all functions $f : X \times X \to \mathbb{F}$, with the property that $f(x, y) \neq 0$ only if $x \leq y$. We equip the set $\mathcal{F}(X)$ with the *convolution product*

$$(f * g)(x, y) = \sum_{x \leq z \leq y} f(x, z) g(z, y).$$

It is straightforward to verify that the convolution product is associative.
We recall from the introduction that the *Kronecker delta function* is defined by $\delta(x, y) = 1$ if and only if $x = y$, and it is zero otherwise. The Kronecker delta function acts as the identity function with respect to the convolution product, since $\delta * f = f * \delta = f$. Another basic function is the *zeta function* defined by $\zeta(x, y) = 1$ if $x \leq y$, and $\zeta(x, y) = 0$ otherwise.

Theorem 1.4.2. *Let $f \in \mathcal{F}(X)$ be a function such that $f(x, x) \neq 0$ for each $x \in X$. Then there is unique function g so that $f * g = g * f = \delta$.*

Proof. We can define g inductively by first letting

$$g(x, x) = 1/f(x, x),$$

and then letting

$$g(x, y) = - \sum_{x \leq z < y} g(x, z) \frac{f(z, y)}{f(y, y)}.$$

Hence $g * f = \delta$ and so g is the *left inverse* of f. Similarly we can define function h which is the *right inverse* of f. By associativity, $g = g*\delta = g*(f*h) = (g*f)*h = \delta*h = h$.

□

Another basic function is the inverse of the zeta function; it is called the *Möbius function* and denoted by $\mu(x, y)$. We immediately have

Exercise 1.4.3. $\mu(x, x) = 1$ and $\mu(x, y) = -\sum_{x \leq z < y} \mu(x, z)$.

Exercise 1.4.4. Prove by induction on $|B - A|$ that the Möbius function of the Boolean poset $(2^n, \subset)$ is given by $\mu(A, B) = (-1)^{|B-A|}$.

Exercise 1.4.5. Let $(X, \leq)$ be an interval in the set of integers. Then $\mu(x, x) = 1$, $\mu(x, x + 1) = -1$ and $\mu(x, y) = 0$ otherwise.

Let us state the *Möbius inversion formula* (MIF).

Theorem 1.4.6. *Let $(X, \leq)$ be a finite poset and let f, g be two functions from X to $\mathbb{F}$. Then $g(x) = \sum_{y \leq x} f(y)$ for all $x \in X$ if and only if $f(x) = \sum_{y \leq x} g(y)\mu(y, x)$ for all $x \in X$.*

Proof.

$$\sum_{y \leq x} g(y)\mu(y, x) = \sum_{y \leq x} \sum_{z \leq y} f(z)\mu(y, x) =$$

$$\sum_{y \leq x} \mu(y, x) \sum_{z \in X} \zeta(z, y)f(z) =$$

$$\sum_{z \in X} \left(\sum_{z \leq y \leq x} \zeta(z, y)\mu(y, x) \right) f(z) =$$

$$\sum_{z \in X} \delta(z, x)f(z) = f(x).$$

$\square$

If $(X, \leq_1)$ and $(Y, \leq_2)$ are two posets then their *direct product* is the poset $(Z, \preceq)$ where $Z = X \times Y = \{(x, y); x \in X, y \in Y\}$ and $(x, y) \preceq (x', y')$ if $x \leq_1 x'$ and $y \leq_2 y'$.

Theorem 1.4.7. *The Möbius function of the direct product of two posets is the product of their Möbius functions: $\mu((x_1, y_1)(x_2, y_2)) = \mu_1(x_1, x_2)\mu_2(y_1, y_2)$.*

Proof. This can be proven in a straightforward way by induction on the number of pairs (u, v) that lie between (x, y) and (x', y'), and using Exercise 1.4.3. $\square$

The following theorem is an easy consequence of Exercise 1.4.3.

Theorem 1.4.8. *Let P be a finite poset and let P' be P with adjoined smallest and largest elements (denoted by 0 and 1). Let c_i be the number of chains in P' between $0, 1$ of length i. Then $\mu(0, 1) = c_0 - c_1 + \cdots$.*

This means that $\mu(0, 1)$ may be interpreted as the *Euler characteristic* of the abstract simplicial complex associated with P' (see Section 5.1).

Applying Theorem 1.4.6 to the Boolean lattice, and using Exercise 1.4.4, we get

Corollary 1.4.9. *Let F be a function on the set 2^S of all subsets of an n-element set S to a field $\mathbb{F}$. For $K \subset S$ let $G(K) = \sum_{L \subset K} F(L)$. Then*

$$F(K) = \sum_{L \subset K} (-1)^{|K-L|} G(L).$$

Corollary 1.4.9 implies a formula which gives a simple exponential algorithm to calculate a permanent.

Corollary 1.4.10. *Let A be an $n \times n$ matrix. Then*

$$\text{Per} A = \sum_{S \subset \{1, \cdots, n\}} (-1)^{n-|S|} \prod_{i=1}^{n} \left(\sum_{j \in S} a_{ij} \right).$$

Next we consider the poset $(X, \preceq)$ where $X = \{1, 2, \cdots, n\}$ and $a \preceq b$ if a divides b.

Theorem 1.4.11. *Let μ be the Möbius function of $(X, \preceq)$. Then $\mu(1,1) = 1$, $\mu(k,n) = \mu(1, \frac{n}{k})$ if k divides n, $\mu(1,n) = (-1)^m$ if n is a product of m distinct primes, and $\mu(1,n) = 0$ otherwise.*

Proof. The first claim is simple. For the second one let $n = p_1^{\alpha_1} \cdots p_k^{\alpha_k}$ be the unique factorization of n into primes. Clearly, we only need to consider the Möbius function of $(X', \preceq)$, where X' consists of all positive integers that divide n. Clearly, $(X', \preceq)$ is the direct product of the linearly ordered sets $(X_i, \preceq)$, $i = 1, \cdots, k$, where $X_i = \{1, p_i, \cdots, p_i^{\alpha_i}\}$. The theorem thus follows from Theorem 1.4.7 and Exercise 1.4.5.

$\square$

The *Euler function* $\phi(n)$ is equal to the number of integers $1 \leq k \leq n$ such that k and n have the greates common divisor equal to 1. For example $\phi(9) = 6$. A classical formula that precedes all general Möbius inversion formulas reads as follows.

Theorem 1.4.12.
$$\phi(n) = n \prod_{p|n} (1 - 1/p),$$

where the product is over all distinct primes p dividing n.

Proof. Let d divide n. Then $\phi(\frac{n}{d})$ equals the number of integers $1 \leq k \leq n$ such that the greatest common divisor of k, n is d, since any such integer k is of the form $k = dk'$ where $1 \leq k' \leq \frac{n}{d}$ and the greatest common divisor of k' and $\frac{n}{d}$ is 1.
We take the function f in the Möbius inversion formula 1.4.6 to be the Euler function ϕ. Let $g(x) = \sum_{y \preceq x} f(y)$. We observe that $g(n) = n$, since for each $1 \leq k \leq n$ there is unique d such that the greatest common divisor of k and n is d. The inversion formula and Theorem 1.4.11 thus give

$$\phi(n) = \sum_{d \preceq n} \mu(d,n) d = \sum_{d \preceq n} \mu(1,d) \frac{n}{d}.$$

Let $p_1, \cdots, p_k$ be the distinct primes that divide n. Theorem 1.4.11 gives

$$\phi(n) = n - \left(\frac{n}{p_1} + \frac{n}{p_2} + \cdots \right) + \left(\frac{n}{p_1 p_2} + \frac{n}{p_1 p_3} + \cdots \right) - \cdots =$$

$$n \prod_{i=1}^{k} (1 - \frac{1}{p_i}).$$

$\square$

Chapter 2

Introduction to Graph Theory

2.1 Basic notions of graph theory

A graph is an ordered pair of sets (V, E) such that E is a subset of the set $\binom{V}{2}$ of unordered pairs of elements of V. The set $V = V(G)$ is the set of *vertices* and $E = E(G)$ is the set of *edges*. The vertices u and v are the *endvertices* of this edge and we also say that u, v are *adjacent* vertices in G.

We say that $G' = (V', E')$ is a *subgraph* of $G = (V, E)$ if $V' \subset V$ and $E' \subset E$. If a subgraph G' contains all the edges induced by a subset V' of vertices, then G' is called an *induced subgraph*. If $V' = V$ then G' is a *spanning subgraph*. If $V' \subset V$ then $G - V'$ is the subgraph of G induced by $V \setminus V'$. If $E' \subset E$ then $G - E' = (V, E \setminus E')$. If $V' = \{v\}$ and $E' = \{e\}$ then we may write $G - v, G - e$ instead of $G - V', G - E'$. Two graphs are *isomorphic* if one may be obtained from the other by renaming the vertices. The *degree* $\deg_G(v)$ of a vertex v is equal to the number of vertices incident with v. Clearly, the sum of all the degrees equals twice the number of edges of the graph. A subset $E' \subset E$ of edges is called *even* if the graph (V, E') has all degrees even (zero is an even number).

A *walk* in a graph is a sequence $v_1, e_1, v_2, e_2, ..., v_i, e_i, v_{i+1}, ..., e_n, v_{n+1}$ such that each v_j is a vertex and each $e_j = v_j v_{j+1}$ is an edge. The vertices v_1 and v_{n+1} are the endvertices of the walk and n is its length. A walk is called a *trail* if all of its edges are distinct, and a *path* if all of its vertices are distinct. Hence a path is necessarily a trail. A walk whose endvertices coincide is called a *closed walk*. The set of the edges of a path whose endvertices coincide is called a *cycle*. A cycle which induces an induced subgraph on its vertices is *induced cycle*.

A graph $G = (V, E)$ is *connected* if it has a path between any pair of vertices. If a graph is not connected, then it is naturally partitioned into maximal connected *components*.

We will use the symbol P_n to denote a path of length n, and C_n to denote a cycle of length n. Moreover, K_n denotes a *complete graph* with n vertices and all possible edges. Hence K_n has $\binom{n}{2}$ edges.

A graph G is a *bipartite graph* if V may be partitioned into vertex classes (or parts) V_1 and V_2, i.e. $V = V_1 \cup V_2$ and V_1, V_2 are disjoint, such that every edge joins a vertex of V_1 to a vertex of V_2:

$$E \subset \{\{v_1, v_2\}; v_1 \in V_1, v_2 \in V_2\}.$$

We denote by $K_{m,n}$ the complete bipartite graph whose vertex classes have m and n vertices. It is not difficult to make the following observation.

Observation 2.1.1. *A graph G is bipartite if and only if it has no cycle of odd length.*

Proof. A subgraph of a bipartite graph is bipartite and no cycle of an odd length is bipartite. This shows the only if part. On the other hand, if G has no cycle of odd length then the following 'greedy' algorithm finds the bipartition of each component: color an arbitrary vertex by 1, all its neighbours by 2, all their neighbours by 1 and so on. $\qquad\square$

A very natural way to describe a graph is to draw it. Figure 2.1 illustrates the basic graphs by pictures.

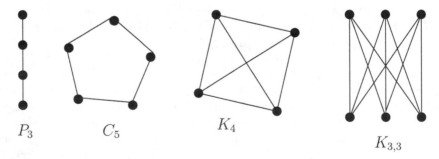

Figure 2.1. Basic pictures of graph theory

A graph with no cycle is a *forest* or an *acyclic graph*. A *tree* is a connected forest. Clearly each forest is bipartite.

Observation 2.1.2. *Each tree with at least one edge has at least two vertices of degree one.*

Proof. The endvertices of any maximal path in the tree are vertices of degree one. $\qquad\square$

A vertex of degree one in a tree is called a *leaf*.

Observation 2.1.3. *For a graph $G = (V, E)$, the following statements are equivalent:*

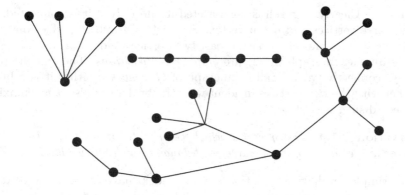

Figure 2.2. Examples of trees

(1) G is a tree,

(2) G is connected and becomes disconnected after deletion of any edge (G is a minimal connected graph),

(3) G is a maximal acyclic graph,

(4) G is connected and $|E| = |V| - 1$.

Proof. All four properties are invariant under deleting or adding a vertex of degree one. We know by Observation 2.1.2 that a tree has a vertex of degree one, so in order to prove the observation by induction on the number of vertices of G it suffices to show that the graphs which satisfy the second or the third or the fourth property must have a vertex of degree one: if G satisfies (2) or (3) then it cannot have a cycle and hence it must have a vertex of degree one by the same argument which proves Observation 2.1.2. If G satisfies (4) then it must have a vertex of degree one since the sum of the degrees equals $2|E|$. $\square$

A tree which is a subgraph of a graph G will be called a *subtree* of G. A subtree is called a *spanning tree* if it contains all vertices of the graph. Similarly a *spanning forest* of a graph is a subforest which contains all vertices. It follows from Observation 2.1.3 that every connected graph has a spanning tree, and every graph has a spanning forest.

If a graph G is connected, and for some set U of *vertices $G - U$* is disconnected, then we say that U *separates* G, or that U is a *vertex cut*, or simply a *cut*. If vertices s, t belong to different components of $G - U$, then we say that U separates s from t. A vertex that separates G is called a *cutvertex*. For $k \geq 2$, we say that G is *k-connected* if either G is a complete graph K_{k+1} or it has at least $k + 2$ vertices and no set of $k - 1$ vertices separates it.

Similarly we can say that a set E' of edges separates a graph G if $G - E'$ is disconnected. An edge that separates G is called a *bridge* of G. A graph is *k-edge-connected* if it has at least two vertices and no set of at most $k-1$ edges that

separates it. Clearly, a graph is 2-connected if and only if it is connected, has
at least three vertices and no cutvertex. Similarly a graph is 2-edge-connected
if and only if it is connected, has at least two vertices and no bridge.

Basic techniques in graph theory are *graph decompositions*. The starting point
is always connectivity. We call a subgraph of G a *block* if either it is a bridge
(together with the two vertices incident with the bridge) or else it is a maximal
2-connected subgraph of G.

Observation 2.1.4. *Every vertex which belongs to at least two blocks is a
cutvertex and conversely every cutvertex belongs to at least two blocks.*

As a simple corollary we get that any two blocks have at most one vertex in
common. We see that a graph decomposes into its blocks: if the blocks of G are
$B_1, \cdots, B_n$ then their edge sets are pairwise disjoint and $E(G) = \bigcup_{i=1}^{n} E(B_i)$.
The decomposition of a connected graph into its blocks has a tree structure: let
$bc(G)$ be the graph whose vertices are the blocks and the cutvertices of G, and
the edges connect each cutvertex to the blocks containing it. Then $bc(G)$ is a
tree. There are several useful tree structures associated with a graph. Let us
call this one *block tree structure*.

Next we study 2-connected graphs. We start with a simple observation which
follows directly from the definitions.

Observation 2.1.5. *If G is 2-connected, then it is 2-edge-connected.*

Theorem 2.1.6. *A graph is 2-connected if and only if there exists, for any pair
of vertices, a cycle containing both of them.*

Proof. The condition is clearly sufficient for the 2-connectivity. Let us prove
the necessity. Let u, v be two vertices of G. We proceed by induction on the
length of the shortest path between u, v. If uv is an edge then the existence of
the cycle containing u, v follows from Observation 2.1.5. Hence let the distance
between u, v be $k > 1$, and let $P = (u = v_1, \cdots, v_{k-1}, v_k = v)$ be a shortest
path connecting u, v. By the induction assumption there is a cycle C containing
u, v_{k-1}. Moreover there is a path Q in $G - v_{k-1}$ connecting u and v. It is not
difficult to observe that $C + Q + \{v_{k-1}, v\}$ contains a cycle with u, v in it (see
Figure 2.3).
$\square$

The operation of *edge subdivision* replaces an edge by a path of length 2
with the same endvertices. Clearly, a graph G is 2-connected if and only if any
graph obtained from G by edge subdivisions is 2-connected. This observation
together with Theorem 2.1.6 implies the following

Corollary 2.1.7. *A graph is 2-connected if and only if there exists, for any
pair of edges, a cycle containing both of them.*

Next theorem shows how each 2-connected graph may be built from smaller
graphs by operations of edge addition and edge subdivision.

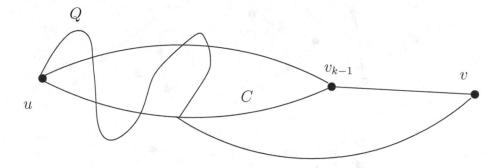

Figure 2.3. Cycles and 2-connectivity

Theorem 2.1.8. *A graph G is 2-connected if and only if it can be constructed from a triangle by a sequence of edge subdivisions and edge additions.*

Proof. Clearly any graph constructed by the two operations is 2-connected. For the other part we first realize that it is the same as saying that each 2-connected graph can be built from a cycle by adding *ears*; an ear is a path that shares only its endvertices with the already constructed part.

By Theorem 2.1.6 every 2-connected graph has a cycle. Now, let $G' \neq G$ be a subgraph of G constructed from a cycle by adding ears. All we have to show is that we can extend it by another ear. If there is an edge not in G' with both endvertices in G' then any such edge forms a new ear. Otherwise, clearly, G has an edge uv where $u \in G'$ and $v \notin G'$. Moreover, $G - u$ has a path connecting v to G' since G is 2-connected. This path together with the edge uv forms the desired new ear.

$\square$

There are several notions closely related to that of a graph. By definition a graph does not contain a *loop*, i.e. an 'edge' joining a vertex to itself; neither does it contain *multiple edges*, that is, several 'edges' joining the same pair of vertices. In a *multigraph* both multiple edges and multiple loops are allowed. In multigraphs we can introduce the operation of *edge contraction* as follows:

Definition 2.1.9. Let $G = (V, E)$ be a multigraph and let e be its edge. Then we denote by G/e the multigraph obtained from G by identification of the endvertices of e. We note that the contraction possibly creates a loop and multiple edges. If $V' \subset V$ then we denote by G/V' the multigraph obtained from G by identification of all the vertices of V' into a single vertex.

The contraction is often applied in the class of simple graphs as well. Clearly, for formal reasons, in the class of simple graphs the multiple edges and the loops must be deleted after the contraction.

If the edges are *ordered* pairs of vertices, then we get the notions of a *directed graph* and a *directed multigraph*. An ordered pair (u, v) is said to be a *directed*

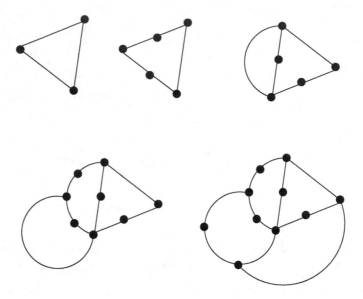

Figure 2.4. Building of a 2-connected graph

edge or an *arc*, and it is *directed from* vertex u to vertex v. A directed walk, directed trail, directed path, directed cycle, indegree, outdegree are notions used in an obvious way. The *underlying multigraph* of a directed graph is obtained by forgetting the orientation of each edge. An *orientation* of a graph G is a directed graph obtained by orienting the edges of G, that is, by giving each edge uv an orientation (uv) or (vu).

2.2   Cycles and Euler's theorem

In this introductory part of the book we are assembling some connections between combinatorics and statistical physics. Perhaps the first theorem of graph theory is Euler's theorem. Let us first write down an observation. A subset of directed edges will be called *even* if for every vertex its indegree equals its outdegree.

Observation 2.2.1. *Each even set of (directed) edges may be partitioned into disjoint (directed) cycles.*

Proof. This observation might be called the *greedy principle of walking.* To prove it we observe first that each non-empty even set contains a cycle since any (directed) path in it may be prolonged; if we delete a (directed) cycle from an even set, we again get an even set, and we can continue in this way until the remaining set is empty. In this way we can construct a desired partition.   □

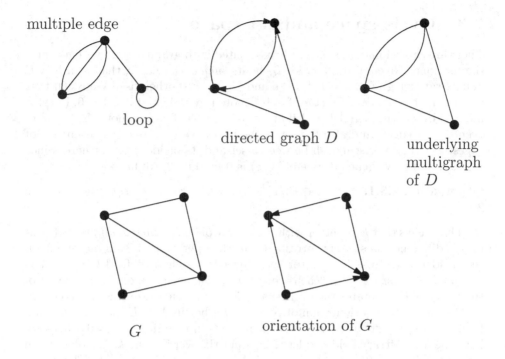

multiple edge

loop

directed graph D

underlying
multigraph
of D

G

orientation of G

Figure 2.5. More concepts of graph theory

A (directed) graph is called *Eulerian* if it has an *Euler tour*, i.e. a closed (directed) trail containing all the (directed) edges. Here comes Euler's theorem.

Theorem 2.2.2. *A (directed) graph is Eulerian if and only if its edge set E is even and the (underlying multigraph of the directed) graph is connected.*

Proof. Follows from Observation 2.2.1 since any connected collection of disjoint (directed) cycles may be combined into a (directed) closed trail. □

A connected graph is Eulerian if one can walk through all its edges exactly once and return back to the origin. Theorem 2.2.2 provides a characterization of Eulerian graphs, and its proof also gives an efficient algorithm for finding an Euler tour. It turns out that the problem changes drastically if we want to walk through *each vertex* exactly once.

In the *traveling salesman problem (TSP)*, a salesman is to make a tour of n cities, at the end of which he has to return to the city he starts from. The cost of the journey between any two cities is known. The TSP asks for (the cost of) a least expensive tour. This basic discrete optimization problem belongs to the class of NP-*complete problems*, where the existence of an efficient algorithm is considered unlikely. If the cost of the journey between a pair of cities is either 1 or ∞, we get the problem of finding a cycle in a graph, that goes through all the vertices. Such a cycle is called a *Hamiltonian cycle*; deciding whether a graph has a Hamiltonian cycle is also an NP-complete problem.

2.3   Cycle space and cut space

There are several matrices naturally associated with graphs. Here we encounter the first one. Given a graph G let I_G be its *incidence matrix*, that is, a $V \times E$ matrix satisfying $(I_G)_{ve} = 1$ if $v \in e$ and $(I_G)_{ve} = 0$ otherwise. We recall that for $A \subset E$, the *incidence vector* of A is denoted by $i(A)$ and it is the $0, 1$ vector indexed by E, where $[i(A)]_e = 1$ if and only if $e \in A$. Let $\mathcal{K}$ denote the set of all even sets of edges, and also the set of the incidence vectors of the even sets of edges: we will not distinguish between a set and its incident vector here. Since $z \in \mathcal{K}$ if and only if each degree of (V, z) is 0 modulo 2, we have

Observation 2.3.1. $\mathcal{K} = \{z \in \{0,1\}^E; I_G z = 0\}$, *where the equality is modulo* 2.

This means that $\mathcal{K}$ together with the operation of *symmetric difference* (symmetric difference on subsets corresponds to the operation of 'sum modulo 2' on the incidence vectors) is a vector space over the 2-element field $GF(2)$. It is called the *cycle space*. Since each even set is a disjoint union of cycles, the set of the cycles of G generates the cycle space. Next we generalize this construction. If D is an arbitrary orientation of G we let I_D be the $V \times E$ matrix satisfying $(I_D)_{ve} = 1$ if e starts in v, $(I_D)_{ve} = -1$ if e ends in v, and $(I_D)_{ve} = 0$ otherwise. Let $\mathbb{F}$ be an arbitrary field and let A be a matrix over $\mathbb{F}$. The *kernel* of A is the set $\{x \in \mathbb{F}^E; Ax = 0\}$ and the *image* of A is the set $\{xA; x \in \mathbb{F}^V\}$. The kernel and the image of a matrix are orthogonal complements of each other.

The kernel of I_D over $\mathbb{F}$ is called the *cycle space* of G over $\mathbb{F}$ and the image of I_D is the *cut space* of G over $\mathbb{F}$. Hence the cycle space and the cut space are orthogonal complements. It is not hard to see that the cycle spaces corresponding to different orientations of G are isomorphic, and the same holds for the cut spaces.

Observation 2.3.2. *The cycle space of G (over a field $\mathbb{F}$) depends only on G, not on the orientation D. The dimension of the cut space (over $\mathbb{F}$) is $|V| - k$ and the dimension of the cycle space (over $\mathbb{F}$) is $|E| - |V| + k$, where k is the number of the connected components of G.*

Proof. Let $T \subset E$ be a maximum acyclic set of edges of G. We learned in the introductory part on trees that $|T| = |V| - k$, and if $e \in E - T$ then $T \cup \{e\}$ contains exactly one cycle which we denote by C^e.

Let C be a cycle of G. We associate to it a vector as follows: Let us choose one of the two ways of walking along C and let us denote by $C(D)$ the vector from $\{0, 1, -1\}^E$ defined by

$C(D)_e = 1$ if $e \in C$ and e is oriented in D in agreement with our walking,

$C(D)_e = -1$ if $e \in C$ and e is oriented in D against our walking,

and $C(D)_e = 0$ otherwise.

We will show that $\{C^e(D); e \in E \setminus T\}$ forms a basis of $\mathcal{K}$ over $\mathbb{F}$. This implies all the claims of the observation. For instance, the dimension of the cut space is $|V| - k$ since the dimensions of the complementary cycle and cut spaces sum up to $|E|$. The vectors $C^e(D)$ are clearly linearly independent since each is nonzero at a unique and different edge of $E \setminus T$. Each element z of the cycle space is a *circulation*, i.e., for each vertex $v \in V$, we have the following equality in $\mathbb{F}$:

$$\sum_{(uv) \in E} z_{(uv)} = \sum_{(vu) \in E} z_{(vu)}.$$

It is not difficult to observe that each circulation is a linear combination of vectors $C(D)$ for some cycles C of G. Hence let us assume that $z = C(D)$ for some cycle C of G. Let $s = \sum_{e \in C-T} \text{sign}(C, D, e) C^e(D)$ where $\text{sign}(C, D, e) = 1$ if $C^e(D)_e = C(D)_e$ and $\text{sign}(C, D, e) = -1$ otherwise. Then $z + (-s)$ is a circulation since both z, s are. Moreover z and s are identical in the entries indexed by $E \setminus T$ and so the circulation $z + (-s)$ is zero in a complement of an acyclic set. But that is possible only if $z + (-s) = 0$, i.e. $z = s$. $\qquad \square$

Definition 2.3.3. A set $E' \subset E$ is an edge cut if there is a partition of V into two sets V_1, V_2 so that $E' = \{\{u, v\} \in E : u \in V_1, v \in V_2\}$.

Example 2.3.4. Each non-empty edge cut separates G, and each separating set of edges contains a non-empty edge cut. The empty set is the edge cut corresponding to the trivial bipartition $(V, \emptyset)$ of V.

Where does the name *cut space* come from? We saw above that the cycle space of a graph G over a field $\mathbb{F}$ is the set of all circulations (over $\mathbb{F}$) with respect to an orientation $D = (V, E)$ of G. Similarly it is not difficult to see that the cut space over $\mathbb{F}$ is the set of all *potential differences* in D; given a function p (a *potential*) on V with values in $\mathbb{F}$, the corresponding potential difference is the function p' on E defined by $p'(uv) = p(u) - p(v)$ for each $(uv) \in E$. If the field $\mathbb{F}$ is $GF(2)$ then the potential differences coincide with the incidence vectors of the *edge cuts*.

Max-Cut and *Min-Cut* problems belong to the basic hard problems of computer science. Given a graph $G = (V, E)$ with a (rational) weight $w(e)$ assigned to each edge $e \in E$, the *Max-Cut problem* asks for the maximum value of $\sum_{e \in E'} w(e)$ over all edge cuts E' of G, while the *Min-Cut problem* asks for the minimum of the same function. The Max-Cut problem is hard (NP-complete) for non-negative edge-weights and hence both Max-Cut and Min-Cut problems are hard for general rational edge-weights. The Min-Cut problem is efficiently (polynomially) solvable for non-negative edge-weights. This is a fundamental result of computer science, known as the max-flow, min-cut algorithm (see Theorem 2.4.1). Still, there are some important classes of graphs where the general Max-Cut problem is efficiently solvable. One such class is the class of the planar graphs.

A basic question about cycles is, *when do two graphs have isomorphic cycle spaces*, i.e. when is there a bijection between their sets of edges that induces a bijection between the sets of their cycles. This we treat next.

Definition 2.3.5. Let $G = (V, E)$ and $G' = (V', E')$ be graphs. We say that G' is *cycle isomorphic* to G if there is a bijection f from E to E' so that z is a cycle of G if and only if $f(z)$ is a cycle of G'.

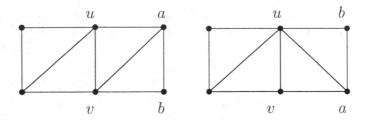

Figure 2.6. An example of a twist

We now review a classical theorem of Whitney. Let G be a 2-connected graph with vertex cut $\{u, v\}$. Let there be subgraphs G_1 and G_2 such that $V(G_1) \cup V(G_2) = V(G)$, $V(G_1) \cap V(G_2) = \{u, v\}$, $E(G_1) \cup E(G_2) = E(G)$ and $E(G_1) \cap E(G_2) = \{uv\}$ or $\emptyset$ depending on whether or not uv is an edge of G. Let G' be the graph obtained from G by replacing each occurrence of u in an edge of G_1 with v and each occurrence of v in an edge of G_1 with u (the neighbors of u and v in G_1 are interchanged). Then G' is obtained from G by a *twist*, and G' is a *twist* of G (see Figure 2.6).

Theorem 2.3.6. *Let G be a 2-connected graph with n vertices and let G' be cycle isomorphic to G. Then G' can be transformed to a graph isomorphic to G by a sequence of at most $n - 2$ twists.*

In order to prove Theorem 2.3.6, we write down a sequence of lemmas.

Lemma 2.3.7. *If G is 3-connected then G' is isomorphic to G.*

Proof. Let $f : E \to E'$ be a cycle isomorphism of G and G'. The cut space is the orthogonal complement of the cycle space and so C is an edge cut of G if and only if $f(C)$ is an edge cut of G'. Further, if $C_1 \cap C_2 = \emptyset$, C_1 is an edge cut and C_2 is a cycle of G, then $f(C_1) \cap f(C_2) = \emptyset$, $f(C_1)$ is an edge cut and $f(C_2)$ is a cycle of G'.

Let v be a vertex of G. Let $N(v)$ denote the set of edges incident to v. Since G is 2-connected, $N(v)$ is a minimal (w.r.t. inclusion) edge cut. Since $G - v$ is 2-connected, $N(v)$ has the property that each pair of edges of $E - N(v)$ belongs to a common cycle of $E - N(v)$ (see Corollary 2.1.7). Further it is not difficult to see that in a 2-connected graph, any minimal edge cut with the above property must be the neighborhood $N(u)$ for some vertex u.

We have that each $f(N(v))$ is a minimal edge cut of G' and each pair of edges of $E' - f(N(v))$ belongs to a common cycle of $E' - f(N(v))$. Further observe that G' is 2-connected since each pair of edges of G' belong to a common cycle (again by Corollary 2.1.7). Hence the edge cuts $f(N(v)), v \in V$ are exactly the edge cuts of G' around the vertices. Cycle isomorphism f thus induces an isomorphism of G and G'.

$\square$

Definition 2.3.8. Let $k \geq 2$. A connected graph G is a *generalized cycle* with parts $G_1, \cdots, G_k$ if the following conditions hold.

(1) Each G_i is connected with at least one edge and if $k = 2$ then both G_1, G_2 have at least three vertices.

(2) Each G_i shares exactly two vertices (called contact vertices) with $\cup_{j \neq i} G_j$.

(3) The edge-sets of G_i's partition E.

(4) If each G_i is replaced by an edge joining its contact vertices then the resulting graph is a cycle.

Lemma 2.3.9. *Let G be 2-connected but not 3-connected and have at least 4 vertices. Then G has a representation as a generalized cycle where each part is 2-connected or an edge.*

Proof. Let $\{u, v\}$ be a cut of G and let H_1 be a component of $G - \{u, v\}$. Let $H_2 = G - \{u, v\} - H_1$. Let G_1 be the subgraph of G induced by $H_1 \cup \{u, v\}$ and let G_2 be the subgraph of G induced by $H_2 \cup \{u, v\}$ where we delete edge $\{u, v\}$ if it belongs to G. Then G is a generalized cycle with parts G_1, G_2. If G_1 is neither an edge nor 2-connected then it has a vertex w distinct from both u and v such that w cuts G_1 into two connected parts $G_{1,1}$ and $G_{1,2}$ and we get a representation of G as a generalized cycle with parts $G_{1,1}, G_{1,2}, G_2$. Clearly we can repeat this procedure until the claim is fulfilled. $\square$

Following notation introduced earlier we say that a graph is a *block* if it is 2-connected, or an edge. Let us call a generalized cycle *robust* if each part is a block. The next lemma is the key to Whitney's theorem (Theorem 2.3.6).

Lemma 2.3.10. *Let G be a graph represented as a robust generalized cycle with parts $G_1, \cdots, G_k$. Let G' be a graph and let $f : E \to E'$ be a cycle isomorphism of G and G'. Then G' is a generalized cycle with parts $G'_1, \cdots, G'_k$ where each G'_i is induced by $f(E(G_i))$.*

Proof. G is 2-connected and thus any pair of its edges belong to a common cycle. Since G' is cycle isomorphic to G, every pair of edges of G' belongs to a common cycle and G' is thus 2-connected as well. Each G'_i is cycle isomorphic to G_i. Hence each G'_i is a block. We let $G_{-i} = \cup_{j \neq i} G_j$, and analogously we define G'_{-i}. Since G and consequently also G' have a cycle containing at least one edge from each part, Lemma 2.3.10 will follow if we show that

$$|V(G'_{-i}) \cap V(G'_i)| = 2.$$

This is clearly true if G'_i is an edge. Hence let G'_i be 2-connected. The existence of a cycle meeting all parts of G' implies that $|V(G'_{-i}) \cap V(G'_i)| \geq 2$.

Let x be a contact vertex of G_i and let E_x be the set of the edges of G_i which contain x. Then $f(E_x)$ is a minimal edge cut in G'_i since E_x is a minimal edge cut in G_i. The key property (a consequence of the analogous property of G_i) is that

Every cycle of G' that contains an edge of G'_i and an edge of G'_{-i} also contains an edge of $f(E_x)$.

Let $V(G'_{-i}) \cap V(G'_i) = C$. Let us assume for a contradiction that C has at least three vertices. Let e^1, e^2, e^3 be three edges of G'_{-i} incident to three different vertices of C. Each pair of e^i, e^j belongs to a common cycle. Hence, there are at least two pairs of vertices from C (say uv and wz), where at least three vertices among u, v, w, z are distinct, so that each pair is connected in G'_{-i} by a path whose inner vertices do not belong to G'_i.

Since $f(E_x)$ is a minimal edge cut in G'_i, the graph $G'_i - f(E_x)$ has exactly two connected components. This together with the existence of the pairs uv and wz produces a cycle in G' which contains an edge of G'_i, an edge of G'_{-i}, but no edge of $f(E_x)$. This contradicts the key property above.

$\square$

The following lemma completes the proof of Theorem 2.3.6. It turns out to be convenient to formulate the lemma for graphs where some edges are directed. An isomorphism of such graphs must keep the orientation of the directed edges.

Lemma 2.3.11. *Let G and G' be 2-connected graphs such that exactly one edge of each graph, $e \in E, e' \in E'$ are directed. Let $f : E \to E'$ be a cycle isomorphism and $f(e) = e'$. Then G' may be transformed by a sequence of at most $|V| - 2$ twists into a graph isomorphic with G by the function f.*

Proof. If $|V| = 2$ or G is 3-connected then the lemma holds (by Lemma 2.3.7; at most one twist to preserve the direction of e is needed).

Now we proceed by induction. Let $|V| = n$, and assume the lemma is true for any graph with at most $n - 1$ vertices. Let G be represented as a robust generalized cycle with parts $G_1, \cdots, G_k$ (numbered along the cycle), and let G' be represented as a robust generalized cycle with parts $G'_1, \cdots, G'_k$ such that $f(E(G_i)) = E(G'_i)$. Let $e \in G_1$. Let u_i be the contact vertex between G_i and G_{i+1} (counting is modulo k). Analogously let u'_i be the contact vertex between G'_i and G'_{i+1}.

Let H_1 (H'_1 respectively) be obtained from G_1 (G'_1 respectively) by adding the edge $\{u_k, u_1\}$ ($\{u'_k, u'_1\}$ respectively). Let us define $f_1 : E(H_1) \to E(H'_1)$ to be equal to f restricted to G_1, and $f(\{u_k, u_1\}) = \{u'_k, u'_1\}$. Then f_1 is a cycle isomorphism and by the induction assumption we can transform H'_1 into a graph isomorphic to H_1 by at most $n_1 - 2$ twists, where n_1 is the number of vertices of H_1.

By these twists G'_1 is transformed into a graph G_1^* isomorphic to G_1; the isomorphism is induced by f_1 and we denote it by F_1. For $i \neq 1$, G'_i is unchanged, but the contact vertices might be interchanged: G'_2 may be attached to G_1^* at $F_1(u_1)$ or at $F_1(u_k)$.

Further, let H_2 be obtained from G_{-1} by adding the *directed edge* (u_1, u_k). Let H'_2 be obtained from G'_{-1} by adding the directed edge $(F_1(u_1), F_1(u_k))$ or $(F_1(u_k), F_1(u_1))$, depending on whether F_1 keeps or interchanges the contact

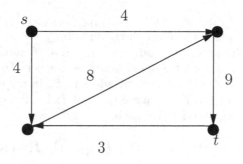

Figure 2.7. A cut of capacity 13

vertices. Analogously as above, applying at most $n - n_1 + 2 - 2$ twists to G'_{-1} we obtain a graph G^* isomorphic to G. We used at most $(n_1 - 2) + (n - n_1) = n - 2$ twists.

$\square$

In the next sections we study graph connectivity and network flows. The flows are flowing as much as we are walking in this book.

2.4 Flows in directed graphs

Let $D = (V, E)$ be a directed graph and let $v \in V$. A function f on E is a *flow at* v if it satisfies Kirchhoff's current law: the total flow into v is equal to the total flow leaving v, or $\sum_{(uv) \in E} f(uv) = \sum_{(vu) \in E} f(vu)$. A *circulation* is a flow at each vertex of D.

We study flows from vertex s (the *source*) to vertex t (the *sink*). We say that a function f on E is an *s,t-flow*, if it is a flow at each vertex different from both s, t. Clearly, whichever flows out of s in an s, t−flow, must flow into t. Hence we may define the *value* $|f|$ of the s,t-flow f by

$$|f| = \sum_{(sv) \in E} f(sv) - \sum_{(us) \in E} f(us) = \sum_{(ut) \in E} f(ut) - \sum_{(tv) \in E} f(tv).$$

The Max-Flow Problem. We are given a directed graph $D = (V, E)$, two different vertices s, t of D and a non-negative rational *capacity* $c(e)$ of each edge $e \in E$. The max-flow problem is to find a maximum value s, t−flow f such that for each $e \in E$, $0 \leq f(e) \leq c(e)$.

If $S \subset V$ such that $s \in S$ and $t \notin S$ then let us denote by S^+ the set of directed edges leaving S, i.e., $S^+ = \{(u, v); u \in S, v \notin S\}$. Each such set S^+ is called an s, t-*cut*, and its capacity $c(S^+)$ is defined by $c(S^+) = \sum_{e \in S^+} c(e)$.

Theorem 2.4.1. *The maximum* s, t−*flow value is equal to the minimum capacity of an* s, t−*cut.*

Proof. We have $|f| = \sum_{(uv)\in S^+} f(uv) - \sum_{(vu)\in E, v\notin S, u\in S} f(vu)$ for each $s,t-$cut S^+ and so the maximum is at most the minimum.

In order to finish the proof, we update an $s,t-$flow f and the subset $S \subset V$ recursively as follows. First let $S = \{s\}$ and f be the zero flow. If $u \in S$, $v \notin S$, and $c(u,v) > f(uv)$ or $f(vu) > 0$ then we add v to S. In the end two things may occur. If $t \in S$ then we can improve the value of f: let $s = v_0, v_1, \cdots, v_l = t$ be a path in S connecting s and t. We have

$$0 < \epsilon_i = \max\{(c(v_i, v_{i+1}) - f(v_i, v_{i+1})), f(v_{i+1}, v_i)\}.$$

Let $\epsilon = \min \epsilon_i, i = 0, \cdots, l - 1$. Then f may be *augmented* along that path to a flow f' with value $|f'| = |f| + \epsilon$ as follows: if $\epsilon_i = c(v_i v_{i+1}) - f(v_i v_{i+1})$ then let $f'(v_i v_{i+1}) = f(v_i v_{i+1}) + \epsilon$, else let $f'(v_{i+1} v_i) = f(v_{i+1} v_i) - \epsilon$.

On the other hand, if $t \notin S$, then f has maximum value since

$$|f| = \sum_{u\in S, v\notin S} f(uv) - \sum_{u\notin S, v\in S} f(uv) = \sum_{u\in S, v\notin S} c(uv) = c(S^+).$$

The capacities are rational and the graph is finite and so in each step we augment the flow by at least a fixed non-zero fraction. Hence a maximum flow is reached in a finite number of steps.

$\square$

We remark that the proof also shows the following

Corollary 2.4.2. *If the capacity function is integral then there is a maximal flow that is integral.*

Let us assume now that we have capacity restrictions on the vertices, except for the source and the sink, and we want to find a maximum value flow f such that for each vertex v,

$$\sum_{(u,v)\in E} f(u,v) \le c(v).$$

This is reduced to flows by an important operation called *splitting*.

Definition 2.4.3. Let $D = (V, E)$ be a directed graph and let $v \in V$. We say that D' is obtained from D by *splitting a vertex v* if v is replaced by two new vertices v_s, v_t so that each edge of D starting in v starts in v_s of D', each edge of D terminating in v terminates in v_t of D', and $e(v) = (v_t, v_s) \in E(D')$.

We say that a set U of vertices is an *s,t-vertex cut* if $D - U$ has no directed path from s to t.

Theorem 2.4.4. *Let D be a directed graph with a rational capacity $c(v) \ge 0$ on each vertex v different from s,t. Then the minimum capacity of an s,t-vertex cut is equal to the maximum s,t-flow value.*

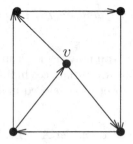 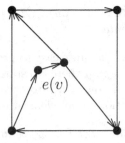

Figure 2.8. Splitting a vertex v

Proof. Let D' be obtained from D by splitting each vertex v different from s, t. We also let $c(v_t v_s) = c(v)$ and $c(e) = +\infty$ for all other edges of D'. Clearly, there is a bijection between the set of the flows of D and the set of the flows of D' and the $s, t-$cuts of D' of finite capacity exactly correspond to the $s, t-$vertex cuts. □

2.5 Connectivity

Let $G = (V, E)$ be a graph and let s, t be two distinct vertices of G. We say that two $s, t-$paths are *independent* if they have only the vertices s, t in common. The following Menger's theorem is perhaps the most important theorem regarding graph connectivity.

Theorem 2.5.1. *Let s and t be distinct non-adjacent vertices of a graph G. Then the minimum number of vertices separating s from t is equal to the maximum number of independent s,t-paths. Moreover, if s, t are distinct vertices of G, then the minimum number of edges separating s from t is equal to the maximum number of edge-disjoint s,t-paths.*

Proof. This follows from the max-flow min-cut theorems 2.4.1, 2.4.2 and 2.4.4 if we replace each edge by two oppositely oriented edges with the same endvertices, and let the capacities of the vertices (for the first statement) and of the edges (for the second statement) be 1. □

Corollary 2.5.2. *For $k \geq 2$, a graph is k-connected if and only if it has at least 2 vertices and any two vertices may be joined by k independent paths. Also, a graph is k-edge-connected if and only if it has at least two vertices and any two vertices may be joined by k edge-disjoint paths.*

The last observation on connectivity characterizes 2-edge-connectivity by the existence of an orientation. The relations of connectivity and the existence of orientations with particular properties is a flourishing research subject with many results much stronger than the one we present below.

We say that a directed graph is *strongly connected* if there is a directed path from each vertex to any other vertex.

Theorem 2.5.3. *A graph is 2-edge-connected if and only if it has a strongly connected orientation.*

This theorem is quite easy to prove directly, you may try it as an exercise. It is a nice surprise that there is a statement about integer lattices hiding behind it. This we explain now. An *integer lattice* is a subset of $\mathbb{Z}^d$ closed under addition and multiplication by an element of $\mathbb{Z}$.

Theorem 2.5.4. *Let $L \subset \mathbb{Z}^d$ be an integer lattice and let $A \subset L$. There exists $s : A \to \{1, -1\}$ such that each element of L is a non-negative integral linear combination of $\{s(a)a : a \in A\}$ if and only if the following two conditions are satisfied:*

(1) Each element of L is an integral linear combination of elements of A.

(2) For each $a \in A$, there are integers $i_a(b), b \in A$, such that $i_a(a) \neq 0$ and $0 = \sum_{b \in A} i_a(b)b$.

Proof. Condition (1) is clearly necessary. To show that Condition (2) is necessary let s exist. Let $a \in A$. Since $-s(a)a \in L$, we have $-s(a)a = \sum_{b \in A} i_a(b)b$ where $i_a(b) = 0$ or $i_a(b)$ has the same sign as $s(b)$. Adding $s(a)a$ to both sides, we get (2). Let us prove that the two conditions are sufficient. In fact, we will prove a stronger statement.
Let us assume that $A \cup \{-a; a \in A\}$ generates L. Moreover let $A' \subset A$ and $s' : A' \to \{1, -1\}$ be given with the following property **P**:
For each $a \in A$ we may write $0 = \sum_{b \in A} i_a(b)b$, where each $i_a(b)$ is integer, $i_a(a) \neq 0$ and for $b \in A'$, if $i_a(b) \neq 0$ then it has the same sign as $s'(b)$.

Claim. Let $b \in A \setminus A'$ arbitrary. Then s' may be extended to $s'' : A' \cup \{b\} \to \{1, -1\}$ so that **P** is valid for $A' \cup \{b\}$ and s''.
    Proof of Claim. For contradiction assume that s' cannot be extended to $A' \cup \{b\}$. This means that, if we let $s''(b) = 1$, then **P** is violated for some $x \in A$, and if we let $s''(b) = -1$, then **P** is violated for some $y \in A$. Since **P** holds for s' we have that $x \neq y$ and $0 = \sum_{c \in A} i_x(c)c$, where $i_x(x) \neq 0$, $i_x(b) < 0$ and for each $c \in A'$, $i_x(c) = 0$ or it has the same sign as $s'(c)$. We also have $i_x(y) = 0$ since otherwise y does not contradict property **P** for $s''(b) = -1$. Analogously for y we have $i_x(y) = 0$ and $0 = \sum_{c \in A} i_y(c)c$, where $i_y(y) \neq 0$, $i_y(b) > 0$ and for each $c \in A'$, $i_y(c) = 0$ or it has the same sign as $s'(c)$. Without loss of generality assume that $-i_x(b) \geq i_y(b)$. We set, for each $c \in A$, $i'_y(c) = i_x(c) + i_y(c)$. Then $i'_y(y) \neq 0$, $i'_y(b) \leq 0$, and $0 = \sum_{c \in A} i'_y(c)c$. Moreover, for each $c \in A'$, $i'_y(c) = 0$ or it has the same sign as $s'(c)$. Hence if we let $s''(b) = -1$ then **P** is satisfied for y and s'', which contradicts the choice of y. $\square$

Corollary 2.5.5. *Let L be an integer lattice and $A \subset L$ such that each element of L is an integral linear combination of $A - \{a\}$, for any $a \in A$. Then a signing s as in Theorem 2.5.4 exists.*

Proof of Theorem 2.5.3. Let $V = \{1, ..., n\}$. For any $i < j$ let $\sigma(i, j) \in \{0, 1, -1\}^n$ be a vector whose components are all equal to zero except $\sigma(i, j)_i = 1$ and $\sigma(i, j)_j = -1$. Let L be the integer lattice generated by all the vectors $\sigma(i, j)$, and let $A = \{\sigma(i, j); \{i, j\} \in E\}$. Then G has a strongly connected orientation if and only if signing s as in Theorem 2.5.4 exists for A. Moreover, the two conditions of Theorem 2.5.4 are equivalent to G being connected (condition (1)) and each edge belonging to a cycle (condition (2)). This is equivalent to G being 2-edge-connected.

Remark 2.5.6. It may be interesting to find other examples of a 'linear connectivity' in the spirit of Theorem 2.5.4.

2.6 Factors, matchings, and dimers

Let $G = (V, E)$ be a graph. A subgraph (V', E') of G is *spanning* if $V = V'$. Spanning subgraphs are sometimes called *factors*. An important tool in studying factors is their *degree sequence*, i.e. the sequence of the vertex-degrees of E'. Let us first characterize the degree sequences of graphs.

Definition 2.6.1. Let $v_1, \ldots, v_n$ be the vertices of a graph G. The sequence $(\deg(v_1), \ldots, \deg(v_n))$ is called the degree sequence or the score of G.

The degree sequence is determined up to ordering. What can be said about a degree sequence? Clearly

$$\sum_{v \in V} \deg(v) = 2|E|,$$

and in particular, each graph has an even number of odd degrees. However, we can say more: the next theorem gives a recursive characterization of degree sequences.

Theorem 2.6.2. *Let $D = (d_1, \ldots, d_n)$ be a sequence of natural numbers, $n > 1$ and $d_1 \leq d_2 \leq \cdots \leq d_n$. Let $D' = (d'_1, \ldots, d'_{n-1})$ be given by $d'_i = d_i$ for $i < n - d_n$ while $d'_i = d_i - 1$ for $i \geq n - d_n$. Then D is a score if and only if D' is a score.*

Proof. If D' is a score and G' is a graph of score D' then a graph G obtained from G' by adding a new vertex v_n with new edges $\{v_n, v_i\}, i = n-1, \ldots, n-d_n$, has score D. For the other implication we need to show:
If D is a score, then there is a graph G with score D such that a vertex v of the largest degree d_n is connected to a set of d_n vertices different from v, whose degree is at least as large as any remaining degree.
If $d_n = n - 1$ then any graph with score D satisfies this, so let $d_n < n - 1$. Let $\mathcal{G}$ be the set of the graphs of score D. For each $G \in \mathcal{G}$ let $j(G)$ be the largest index so that $\{v_n, v_j\} \notin E(G)$. Let G_0 be a graph such that $j(G_0)$ is as small as possible. We show that $j(G_0) = n - d_n - 1$: this will finish the proof of the theorem. For contradiction let $j > n - d_n - 1$ and suppose that v_j is not joined

by an edge to v_n in G_0. Then there is $i < j$ such that v_i is joined to v_n. There is also $k \neq i, j, n$ such that v_k is joined to v_j but not to v_i since $d_j \geq d_i$ (see Figure 2.9). In this situation we can construct a new graph G' from G by *exchanging*

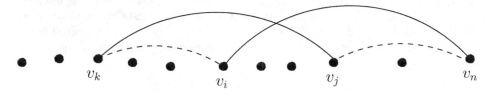

Figure 2.9. An exchange

the edges $v_n v_i, v_k v_j$ for the edges $v_n v_j, v_k v_i$. The graph G' clearly has score D and $j(G') < j = j(G_0)$, which contradicts the choice of G_0.

$\square$

The most interesting factors are the *matchings*. They are the factors which have degree at most one at each vertex. Hence a *matching* in a graph is a set of disjoint edges. A matching is *maximum* if its cardinality is as large as possible, and it is *perfect* if it *covers* all the vertices of the graph, i.e., if its cardinality is half the number of vertices of the graph. In statistical physics, a matching is called a *monomer-dimer arrangement* and a perfect matching is a *dimer arrangement*.

Let $G = (V, E)$ be a graph. We define the *neighbourhood* of $S \subset V$ by $N'_G(S) = \{v \in G - S$; there is $u \in S$ such that $uv \in E\}$. We write $N'(S)$ instead of $N'_G(S)$ if no confusion may arise. If v is a vertex, e is an edge, and $v \in e$ then we say that e *covers* v and v *covers* e.

We start with matchings in bipartite graphs. Let $G = (V, E)$ be a bipartite graph and let V_1, V_2 be its vertex classes. A matching of G is called a *complete matching from V_1* if it covers all vertices of V_1. The following famous theorem of Hall (Theorem 2.6.3) characterizes bipartite graphs with a complete matching. It follows directly from Menger's theorem Theorem 2.5.1.

Theorem 2.6.3. *A bipartite graph G with vertex sets V_1 and V_2 has a complete matching from V_1 if and only if*

$$|N'_G(S)| \geq |S|$$

for every $S \subset V_1$.

Proof. Let G' be obtained from G by introducing two new vertices s, t and by joining s to all vertices of V_1 and t to all vertices of V_2. Menger's theorem applied to s and t says the following: if G does not have a complete matching from V_1 then there are $T_1 \subset V_1$ and $T_2 \subset V_2$ such that $|T_1| + |T_2| < |V_1|$ and there is no edge from $V_1 - T_1$ to $V_2 - T_2$. Then

$$|N'_G(V_1 - T_1)| \leq |T_2| < |V_1| - |T_1| = |V_1 - T_1|.$$

This shows the sufficiency of the condition. The necessity is obvious.   □

Corollary 2.6.4. *If a bipartite graph G with vertex classes V_1, V_2 satisfies*

$$|N_G(S)| \geq |S| - d$$

for every $S \subset V_1$, then G contains a matching of cardinality $|V_1| - d$.

Proof. It suffices to apply Hall's theorem (Theorem 2.6.3) to the graph obtained from G by adding d vertices to V_2 and joining them to each vertex of V_1.

□

Menger's theorem also implies Kőnig's theorem:

Theorem 2.6.5. *Let G be a bipartite graph. The maximum cardinality of a matching is equal to the minimum number of vertices necessary to cover all the edges.*

Bipartite graphs are often successfully used to encode families of sets and $0, 1$ matrices. Given a family $\mathcal{A} = \{A_1, \cdots, A_m\}$ of subsets of a set X, we can form its *incidence bipartite graph* as the bipartite graph $(X \cup \{1, \cdots, m\}, E)$ where $xi \in E$ if and only if $x \in A_i$. We say that $Y \subset X$ is a *system of distinct representatives* for $\mathcal{A}$ if the elements of Y may be numbered so that the i-th element belongs to A_i, $i = 1, \cdots, m$. In the language of the families of sets, Hall's theorem says the following:

Theorem 2.6.6. *A family $\mathcal{A} = \{A_1, \cdots, A_m\}$ of subsets of a set X has a system of distinct representatives if and only if*

$$\left| \bigcup_{i \in F} A_i \right| \geq |F|$$

for every $F \subset \{1, \cdots, m\}$.

Given a general matrix A where the rows are indexed by a set V_1 and the columns are indexed by a set V_2, we can form its bipartite graph $(V_1 \cup V_2, E)$ where $ij \in E$ if and only if $A_{ij} = 1$. Let us call the rows and the columns of A the *lines* of A. Finally we say that a set T of nonzero entries of a matrix is a *transversal* if no line contains more than one element of T. In the language of matrices, Kőnig's theorem says

Theorem 2.6.7. *Let A be a general matrix. Then the maximum cardinality of a transversal of A is equal to the minimum number of lines of A that cover all the non-zero entries.*

Next we move to matchings in general graphs and present a theorem of Tutte which characterizes the graphs with a perfect matching.

Theorem 2.6.8. *A graph $G = (V, E)$ has a perfect matching if and only if for each $S \subset V$, the number of components of $G - S$ of odd cardinality is at most $|S|$.*

Proof. The condition is clearly necessary, so let us prove its sufficiency. Let $o(G')$ denote the number of components of G' of odd cardinality.

If we apply the condition of the theorem for $S = \emptyset$, we obtain that G must have an even number of vertices. Let S_0 be a maximal nonempty set for which the condition of the theorem is satisfied with equality; such a set exists since e.g. each one-element set necessarily satisfies the condition with equality. Let $|S_0| = m > 0$, let $C_1, \cdots, C_m$ be the components of $G - S_0$ of odd cardinality and let $D_1, \cdots, D_k$ be the components of $G - S_0$ of even cardinality.

- *Each D_j has a perfect matching:* For each $S \subset V(D_j)$ we have

$$o(G - S_0) + o(D_j - S) = o(G - (S_0 \cup S)) \leq |S_0| + |S|.$$

  Since $o(G - S_0) = |S_0|$, we have that $o(D_j - S) \leq |S|$. Hence the assertion holds for D_j by the induction assumption.

- *If $v \in C_i$ then $C_i - v$ has a perfect matching:* If not then the condition of the theorem is not satisfied for a set $S \subset V(C_i) - v$ and since $o(C_i - (\{v\} \cup S)) + |\{v\} \cup S|$ has the same (odd) parity as $|V(C_i)|$, we get

$$o(C_i - (\{v\} \cup S)) \geq |S| + 2.$$

  Consequently,

$$|S_0 \cup S \cup \{v\}| \geq o(G - (S_0 \cup S \cup \{v\})) =$$

$$o(G - S_0) - 1 + o(C_i - (S \cup \{v\})) \geq |S_0| + |S| + 1.$$

  Hence $S_0 \cup S \cup \{v\}$ satisfies the condition of the theorem with equality, which contradicts the maximality of S_0.

- *G contains a matching covering S_0 so that the other endvertices of its edges lie in pairwise different components C_j:* We form a bipartite graph with the vertex classes S_0 and $\{C_1, \cdots, C_m\}$ so that C_i is joined to $v \in S$ if there is an edge in G from v to a vertex of C_i. The assertion now follows from Hall's theorem.

The three claims we proved construct a perfect matching of G.

$\square$

Corollary 2.6.9. *A graph has a matching covering all but at most d vertices if and only if for each $S \subset V$, the number of components of $G - S$ of an odd cardinality is at most $|S| + d$.*

A graph is factor-critical if it has a perfect matching after deletion of an arbitrary vertex.

The proof of Theorem 2.6.8 proves more: it demonstrates a useful graph decomposition, the *Edmonds-Gallai decomposition*.

Corollary 2.6.10. *Let $G = (V, E)$ be a graph. Let C be the set of the vertices of G not covered by at least one maximum matching. Let $S \subset V - C$ be the set of the neighbours of the vertices of C, and let D be the set of the remaining vertices. Then each component of C is factor-critical and each component of D has a perfect matching. Furthermore, each maximum matching of G consists of a perfect matching of D, a near perfect matching of each component of C and a matching of S to C so that each vertex of S is matched to a different component of C.*

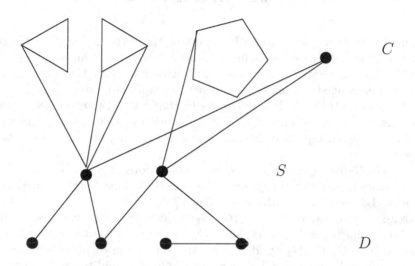

Figure 2.10. Edmonds-Gallai decomposition

Next we turn our attention to the question, when two graphs have the same set of perfect matchings? We recall that the pairs of graphs with the same set of cycles are characterized by Whitney's theorem (Theorem 2.3.6).

Quite surprisingly, the problem for perfect matchings was solved as late as 2004, and only for bipartite graphs. The question for general graphs remains open. A bijection $\psi : E(G) \to E(G')$ is *matching preserving* provided that M is a perfect matching in G if and only if $\psi(M)$ is a perfect matching in G'. For a positive integer k, a graph G is *k-extendable* if G has a matching of size k and every matching in G of size at most k can be extended to a perfect matching in G.

The characterization of the matching preserving bijections for bipartite graphs starts with the twist, the basic operation that preserves cycles (see Figure 2.6). Let G be a connected, 1-extendable, bipartite graph with parts U and V of size n. Let u and v be vertices belonging to different parts of G such that $\{u, v\}$ forms a cut of G. Thus there are bipartite graphs G_1 with parts $U_1 \subseteq U, V_1 \subseteq V$, and G_2 with parts $U_2 \subseteq U, V_2 \subseteq V$, such that $U_1 \cap U_2 - \{u\}$ and $V_1 \cap V_2 - \{v\}$ and each edge of G belongs to either G_1 or G_2 (if uv is an edge of G, then uv is the only common edge of G_1 and G_2). Let G' be the bipartite graph

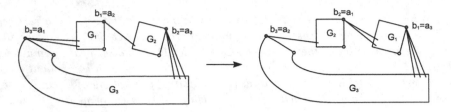

Figure 2.11. Bi-transposition

obtained from G by the *twist* with respect to the vertices u and v. We recall
that this twist consists in replacing each occurrence of u in an edge of G_1 with
v and each occurrence of v in an edge of G_1 with u (the neighbors of u and v in
G_1 are interchanged). Then G' is a bipartite graph with parts $V_1 \cup (U_2 \setminus \{u\})$
and $U_1 \cup (V_2 \setminus \{v\})$. It is easy to verify that these twists preserve perfect
matchings. Are twists sufficient for a description of the matching preserving
bijections between bipartite graphs? The answer is no, one more operation is
needed:
Let G_1, G_2, G_3 be bipartite graphs with bipartitions (V_1^i, V_2^i), $i = 1, 2, 3$, and
having pairwise disjoint vertex sets. We further assume that $|V_1^i| = |V_2^i| + 1$.
Let a_i, b_i be vertices from the same part V_1^i of G_i, $i = 1, 2, 3$. Let G be the
bipartite graph obtained from G_1, G_2, G_3 by identifying the vertices in each of
the three pairs $\{b_1, a_2\}$, $\{b_2, a_3\}$, and $\{b_3, a_1\}$. Let G' be the bipartite graph
obtained from G_1, G_2, G_3 by identifying the vertices in each of the three pairs
$\{b_1, a_3\}$, $\{b_2, a_1\}$, and $\{b_3, a_2\}$. Then the graph G' is said to be obtained from
G by a *bi-transposition* of G_1 and G_2 (see Figure 2.6). It is straightforward to
verify that the operation of bi-transposition also preserves perfect matchings,
and cannot be realised by twists. The following theorem is proved in [BLP].

Theorem 2.6.11. *Let G and G' be two 1-extendable, bipartite graphs and let
$\psi : E(G) \to E(G')$ be a matching preserver. Then there is a sequence of bi-
twists and bi-transpositions of G resulting in a graph isomorphic to G' and ψ is
induced by this isomorphism.*

We will not include a proof of Theorem 2.6.11 here. But, we will explain an
idea that leads to a proof.
There is a useful correspondence between the perfect matchings in a bipartite
graph G with the bipartition U, V and the directed circuits in a directed graph
$D(G, M)$ constructed from G and a specified perfect matching M. Let $D(G, M)$
be the digraph obtained from G by orienting each edge from U to V, and then
contracting all of the edges of M. Let M' be another perfect matching in G.
Then $(M \setminus M') \cup (M' \setminus M)$ is a collection of pairwise vertex disjoint cycles of G of
even length whose edges alternate between M and M'. In $D(G, M)$ these cycles
correspond to vertex-disjoint directed cycles. Using the matching M, we may
reverse this construction to obtain, given a collection of vertex-disjoint directed
cycles of $D(G, M)$, a perfect matching M' of G.

There is a one-to-one correspondence between perfect matchings in G and collections of vertex-disjoint directed cycles in $D(G, M)$.

Further, it is not difficult to observe that G is 1-extendable if and only if $D(G, M)$ is strongly connected. By this transformation we arrive at the problem of when two strongly connected digraphs have the same collections of vertex disjoint directed cycles. We will not discuss this problem here, for its solution see [BLP].

Instead, let us consider a more basic problem, namely when they have the same directed cycles. An isomorphism, respectively, an anti-isomorphism, of a digraph D onto a digraph D' is a bijection $f : V(D) \rightarrow V(D')$ such that, for all $u, v \in V(D)$, there is an arc in D from vertex u to vertex v if and only if there is an arc in D' from vertex $f(u)$ to vertex $f(v)$ (from $f(v)$ to $f(u)$ respectively). A directed twist of a digraph D is defined in a similar way as a twist in a graph. Let D_1, D_2 be subgraphs of D with at least 3 vertices, such that $V(D_1) \cup V(D_2) = V(D)$, $V(D_1) \cap V(D_2) = \{u, v\}$, $E(D_1) \cup E(D_2) = E(D)$. Let D' be obtained from D by replacing arcs of the form uw, wu, vw, and wv by, respectively, wv, vw, wu and uw for each $w \in V(D_2)$, and then reversing the direction of all the remaining arcs of D_2. Then D' is obtained from D by a *di-twist*. Clearly, D and D' have the same circuits and D is strongly connected if and only if D' is. If D is a digraph, then G_D denotes the *underlying graph* of G. Here comes the theorem of Thomassen (see [TC] for the proof) which is the starting point for proving Theorem 2.6.11.

Theorem 2.6.12. *Let D and D' be two strongly connected digraphs with 2-connected underlying graphs G_D and $G_{D'}$. Let $\varphi : E(D) \rightarrow E(D')$ be a bijection such that φ and φ^{-1} preserve directed cycles. Then there exists a sequence of di-twists of D resulting in a digraph D^* such that φ is induced by an isomorphism or anti-isomorphism of D^* onto D'.*

The *matching problem* asks for (the size of) a matching of maximum size, or, when there are weights on the edges, for a matching of the maximum total weight. Next we describe a result of Edmonds who designed a polynomial algorithm for the maximum matching problem in 1963 (appeared in [E1]) and argued that efficient algorithm should be defined by polynomiality. This work is a cornerstone in the development of computer science. In his work on good characterizations ([E2]) the understanding of a nondeterministic algorithm appears. Edmond's solution of the weighted maximum matching problem is a founding result of polyhedral combinatorics, and by conjecturing that there is no polynomial algorithm for the travelling salesman problem he in fact conjectured the famous $P \neq NP$.

Let G be a graph, M a matching in G. A trail is *alternating* if it contains alternately edges of M and out of M. The basic trick for the algorithm is strikingly simple: if M is not a maximum matching, then let M' be a bigger matching (we know it exists but we do not have it in our hands). The symmetric difference of M and M' consists of vertex-disjoint alternating paths and cycles.

Since M' is bigger, there must be an alternating path that starts and ends by an edge out of M. Hence, we get the following.

Observation 2.6.13. *M is maximum if and only if there is no alternating path with both endvertices not covered by M.*

Such a path will be called an *augmenting path*. Let M be a matching. An alternating trail $W = (v_0, e_0, v_1, \cdots, v_t)$ is called a *flower of M* if t is odd, $v_0, \cdots, v_{t-1}$ are all distinct, v_0 not covered by M and $v_t = v_i$ for some *even i*, $0 \le i < t$. Moreover, a cycle C of odd length is called a *blossom of M* if M induces a maximum matching of C, and no edge of M contains only one vertex of C. We note that each flower of M induces a blossom of M' where M' is obtained from M by alternating along $e_0, \cdots, e_{i-1}$. The matchings M and M' have the same size.

Edmonds noticed that the blossoms may be contracted; we recall that to contract a set V' of vertices means to identify them into one vertex, and keep all the edges (since loops are irrelevant when matching is considered, we also delete the loops which appear). The resulting graph is denoted by G/V'.

Observation 2.6.14. *Let B be a blossom of M in G. Then M is a maximum matching in G if and only if $M \setminus B$ is a maximum matching in G/B.*

Observation 2.6.14 naturally leads to a polynomial algorithm. We start with a matching M of G. We construct, in a greedy way, a maximal forest F so that each component has a vertex uncovered by M (its *root*), and each maximal path from a root is alternating and ends with an edge of M. If there is an edge in G between two vertices with an even distance to a (possibly different) root then there is an augmenting path, or a flower and thus a blossom of a matching of the same size as M. We can contract the blossom and continue with the smaller graph. If no such edge exists then the vertices of F of an odd distance to a root prove that M is maximum, by Corollary 2.6.9.

2.7   Graph colorings

The most famous problem-theorem of graph theory is the Four Color theorem (4CT), which may be stated as follows: *The vertices of any planar graph, i.e. any graph that can be drawn in the plane without edge-crossings, may be assigned one of four colors so that no edge has its vertices of the same color.*

This theorem still has only a computer-assisted proof. It is much easier to show that five colors suffice (see Theorem 2.10.4). We will discuss planar graphs including the Four Color theorem in more detail in section 2.10.

Let $G = (V, E)$ be a graph. A *proper k-coloring* of G is any function $f : V \to \{0, \dots, k-1\}$ so that if $uv \in E$ then $f(u) \ne f(v)$. The *chromatic number* $\chi(G)$ is the minimum k so that G has a proper k-coloring.

For instance, $\chi(C_{2k+1}) = 3$, $\chi(K_n) = n$, the chromatic number of a bipartite graph is at most 2 and the 4CT is equivalent to saying that the chromatic number of any planar graph (see 2.10) is at most 4.

Let us denote by $\omega(G)$ the maximum number of vertices of a complete subgraph of G (such a subgraph is called a *clique*), and by $\alpha(G)$ the *independence number* of G, i.e. the maximum number of vertices with no edge between any pair of them. We can start with the following simple observation.

Observation 2.7.1.

$$\chi(G) \geq \max(\omega(G), |V(G)|/\alpha(G)).$$

A very straightforward way to color a graph is to first order the vertices, and then proceed gradually from vertex to vertex and always use the lowest available color. This algorithm may be very bad, as shows the example below; it is a homework to guess a bad ordering there. But on the other hand, there is always an ordering of the vertices so that the greedy algorithm applied to this ordering finds an optimal proper coloring.

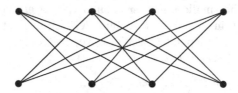

Figure 2.12. Greedy may need $|V|/2$ colors

The greedy algorithm clearly needs at most $\Delta(G) + 1$ colors, where $\Delta(G)$ denotes the maximum degree of G. The next theorem due to Brooks improves this by 1 for most of the graphs.

Theorem 2.7.2. *Let $G = (V, E)$ be a connected graph with maximum degree Δ which is neither a complete graph nor a cycle of odd length. Then $\chi(G) \leq \Delta$.*

A proof may be found in [BB]. Another way to determine the chromatic number of a graph is to count, for each k, the number $M(G, k)$ of proper k-colorings. $M(G, x)$ is known as the *chromatic polynomial*. We will speak more about it in Section 7.2.

We can color edges as well as vertices. A *proper edge-coloring* is a coloring where the edges incident to the same vertex get different colors. The *edge-chromatic number* $\chi'(G)$ is the minimum number of colors in a proper edge-coloring. There is a theorem of Vizing:

Theorem 2.7.3. *A graph of maximum degree Δ has edge-chromatic number Δ or $\Delta + 1$.*

A proof may be found in [BB]. Let us mention that it is algorithmically hard to say, for a given graph, where the truth is.

2.8   Random graphs and Ramsey theory

The $G(n,p)$ model of a random graph is perhaps the best known random graph model. It was invented by Erdős and Renyi. The ground set is the set of all graphs with vertex set $\{1,\cdots,n\}$. To get a random element of the space $G(n,p)$, we select the edges independently, with probability p. Hence the probability of a graph H with m edges is $p^m(1-p)^{\binom{n}{2}-m}$. We are almost always interested in what happens when n goes to infinity, and so it is important to have in mind that p may depend on n: $p = p(n)$. It is customary to denote by $G_{n,p}$ a random element from $G(n,p)$. Next we present a classical example of the use of the random graphs method, the lower bound of Erdős on Ramsey numbers. In fact it gives us a good reason to introduce the basics of Ramsey theory, another milestone of combinatorics. Ramsey theory extends the following principle, known as the pigeon hole principle: if many objects are partitioned into few classes then at least one class is large.

It is convenient to describe the partitions into two parts of the pairs of a finite set as the colorings of the edges of a complete graph by two colors.

Let $R(s,t)$ denote the smallest integer n such that every red-blue coloring of the edges of K_n contains a red K_s or a blue K_t. Equivalently, every graph on n vertices contains an independent set of s vertices of a complete graph of t vertices. Obviously $R(s,t) = R(t,s)$ and $R(s,2) = s$.

Theorem 2.8.1. *If $s > 2, t > 2$, then*

$$R(s,t) \leq R(s-1,t) + R(s,t-1),$$

and

$$R(s,t) \leq \binom{s+t-2}{s-1}.$$

Proof. Let $s + t$ be smallest such that the recursion does not hold. Let $n = R(s-1,t) + R(s,t-1)$; note that n is well defined. Consider a red-blue coloring of the edges of K_n. The degree of each vertex is $n - 1$. A vertex v of K_n thus has at least $n_1 = R(s-1,t)$ red incident edges, or at least $n_2 = R(s,t-1)$ blue incident edges; w.l.o.g. assume the first case. Let K_{n_1} be the complete subgraph of K_n formed by the vertices joined to v by the red edges. K_{n_1} contains either a blue K_t or a red K_{s-1}, and K_n thus has either a blue K_t or a red K_s. This contradicts the choice of n.

The inequality of the theorem holds with equality for $s = 2$ or $t = 2$. For the remaining cases it immediately follows by induction, using the recurrence. □

Let $R(s) = R(s,s)$. The above theorem gives

$$R(s) \leq \frac{2^{2s-2}}{(\pi s)^{1/2}}.$$

The following exponential lower bound has been proved by Erdős, using the random graphs method. No constructive proof is available.

Theorem 2.8.2. *If* $3 \leq s \leq n$ *and*

$$\binom{n}{s} < 2^{\binom{s}{2}-1},$$

then $R(s) \geq n+1$. *Also,*

$$R(s) > 2^{s/2}.$$

Proof. Let $G = (V, E)$ be a graph with n vertices. Let $X(G, s)$ denote the number of complete subgraphs of G of s vertices, and let $X'(G, s)$ denote the number of independent sets of s vertices in G. For the expectation $\mathbb{E}(X(G, s) + X'(G, s))$ in the space $G(n, 1/2)$ we have

$$\mathbb{E}(X(G, s) + X'(G, s)) = 2\binom{n}{s} 2^{-\binom{s}{2}} < 1,$$

and so there is a graph G on n vertices with $X(G, s) + X'(G, s) = 0$. To show the second inequality we observe that $n = n(s) = 2^{s/2}$ satisfies the assumption of the theorem.

$\square$

2.9 Regularity lemma

A set A of positive integers is said to have *positive upper density* if

$$\limsup_{n \to \infty} |A \cap \{1, \cdots, n\}| > 0.$$

In 1975 Szemeredi proved that every set of natural numbers with positive upper density contains arbitrary long arithmetic progressions. A crucial step in the proof is a 'uniformity-regularity' lemma which we now formulate. Roughly speaking, the lemma claims that every graph can be partitioned into a bounded number of almost equal classes such that most pairs of the classes are uniform in the following sense.

For disjoint $U, W \subset V$ let $e(U, W)$ be the number of edges between U, W, and set

$$d(U, W) = \frac{e(U, W)}{|U||W|}.$$

A pair U, W is *ε-uniform* if

$$|d(U^*, W^*) - d(U, W)| < \varepsilon$$

whenever $U^* \subset U$, $|U^*| \geq \varepsilon |U| > 0$ and $W^* \subset W$, $|W^*| \geq \varepsilon |W| > 0$.
A partition $C_0 \cup C_1 \cup \cdots \cup C_k$ of V is *equitable* with exceptional class C_0 if $|C_1| = |C_2| = \cdots = |C_k|$. Finally, an ε-*uniform partition* is an equitable partition such that the exceptional class C_0 has at most εn vertices and all but at most εk^2 pairs (C_i, C_j), $i, j = 1, \cdots, k$, are ε-uniform.

Theorem 2.9.1. *For any positive integer m and every $0 < \varepsilon < 1/2$ there is an integer $M = M(m, \varepsilon)$ such that every graph with at least m vertices has an ε-uniform partition into k parts with $m \leq k \leq M$.*

The proof can be found in [BB]. It seems that the regularity lemma is a general mathematical principle which appears in many diverse considerations. Probably many readers can find its footprint in their own world. Here we include a classical combinatorial consequence (see [BB] for a proof).

Theorem 2.9.2. *For every $\epsilon > 0$ and every graph H, there is a constant $n_0(\epsilon, H)$ with the following property. Let G be a graph with $n \geq n_0(\epsilon, H)$ vertices that does not have H as a subgraph. Then G contains a set E' of less than ϵn^2 edges such that $G - E'$ has no K_r, where $r = \chi(H)$.*

By the way, what is the maximum number of edges of a graph with n vertices and no K_r as a subgraph? By a fundamental theorem of Turan, it is about $(1 - 1/r)\binom{n}{2}$. This is the starting theorem of the extensively studied *Turan-type problems*, and *extremal graph theory*.

2.10   Planar graphs

A lot of results described in this book depend on a representation of a graph on a 2-dimensional surface. In this introductory part we describe some results on *planar representations*, i.e. drawings (embeddings) of graphs in the plane so that the vertices are represented by distinct points, each edge is represented by a curve between the representations of the end-vertices of the edge, and the interior of each edge-representation is disjoint with the rest of the graph representation (i.e., in particular the edges do not cross each other). The graphs that can be represented (embedded) in the plane $\mathbb{R}^2$ in this way are called *planar*, and graphs represented (embedded) in the plane in this way are called *plane* graphs, or *topological planar* graphs. In this section we will deal with the planar embeddings in an intuitive way. More background on embeddings is given in Chapter 5; here we only include the basic Jordan curve theorem (for the proof see [MT]).

Theorem 2.10.1. *Any simple closed curve C in the plane divides the plane into exactly two connected components. Both of these regions have C as the boundary.*

Each plane graph is by definition a subset of the plane. A *face* of a plane graph is any connected component of its planar complement. Exactly one face of a plane graph G is unbounded; it is called the *outer* face of G. For a planar graph G we denote by v, e, p its number of vertices, edges and faces. These numbers are interconnected by the celebrated Euler's formula.

Theorem 2.10.2. *Let G be a connected plane graph. Then*

$$v - e + p = 2.$$

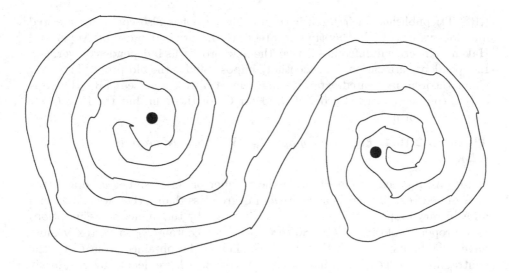

Figure 2.13. The same side?

Proof. We proceed by induction on $e - v$. If G is a tree then $e - v = 1$ and $p = 1$, hence the theorem holds. For the induction step we observe that adding an edge to a connected plane graph increases the number of faces by one.

□

What is the maximum number of edges of a planar graph with v vertices? Quite intuitively the maximum is achieved for a plane graph where each face is a triangle. For such plane graphs we get from Euler's formula $v - e + p = v - e + 2e/3 = 2$, and thus $e = 3v - 6$. Each planar graph with n vertices hence has at most $3n - 6$ edges. In particular, each planar graph has a vertex of degree at most 5. Which graphs are not planar? We leave as an exercise to show (using the Jordan Curve Theorem 2.10.1) that neither K_5 nor $K_{3,3}$ are planar. The Kuratowski's theorem (Theorem 2.10.15) says that these are in fact the only essential non-planar graphs.

Perhaps the most famous theorem of the graph theory is the Four Color theorem.

Theorem 2.10.3. *Each planar graph can be properly colored by 4 colors.*

The Four Color problem dates back to 1852 when Francis Guthrie, while trying to color the map of the counties of England, noticed that four colors sufficed. He asked his brother Frederick if it was true that any map can be colored using four colors in such a way that adjacent regions (i.e., those sharing a common boundary segment, not just a point) receive different colors. Frederick Guthrie communicated the conjecture to DeMorgan, and the rich mathematical history of the problem started ([TIL]). In 1977, Appel and Halton ([AH]) published their computer assisted proof of Theorem 2.10.3, but some scepticism remained regarding its validity. Robertson, Sanders, Seymour and Thomas

([RSST]) published their proof in 1997. This proof is also computer assisted and follows essentially the same strategy as the original proof of Appel and Haken. The crucial difference is that the new proof was independently verified. Independent verification was essentially impossible for the old proof. The Four Color theorem motivated extensive mathematical research, see [TR]. There is only a computer-assisted proof of the Four Color theorem, but the Five Color theorem is not hard to prove.

Theorem 2.10.4. *Each plane graph $G = (V, E)$ can be properly colored with 5 colors.*

Proof. We proceed by induction on the number of vertices. Let u be a vertex of G of degree 5 (if there is a vertex u of degree less than 5 then we can extend an arbitrary proper 5-coloring of $G - u$ (it exists by the induction assumption) to a proper 5-coloring of G). Since G does not contain K_5, a vertex u must have neighbours x, y such that $xy \notin E$. Let G' be obtained from G by the contraction of edges ux, uy into a single vertex which we denote by z. Clearly G' is a plane graph with a smaller number of vertices than G and hence we can assume (by the induction assumption) that G' has a proper 5-coloring. We assume that vertex u has degree 5 in G, and so vertex z has degree 3 in G'. A proper 5-coloring of G' can thus be extended to G so that x and y get the color of the contracted vertex z, and one color remains for u. □

A curve in the plane is a *polygonal arc* if it is the union of a finite number of straight line segments. The following lemma is very intuitive.

Lemma 2.10.5. *Every planar graph may be embedded into the plane so that all edges are represented by simple polygonal arcs.*

An important concept we will need is that of the *dual graph G^** of a plane graph G. It turns out to be convenient to define G^* as an abstract (not topological) graph. But we need to allow multiple edges and loops which is not included in the concept of the graph as a pair (V, E), where $E \subset \binom{V}{2}$. A standard way out is to define a graph as a triple (V, E, g) where V, E are sets and g is a function from E to $\binom{V}{2} \cup V$ which gives to each edge its endvertices. For instance, $e \in E$ is a loop if and only if $g(e) \in V$. Now we can define G^* as the triple $(F(G), \{e^*; e \in E(G)\}, g)$ where $F(G)$ is the set of the faces of G and $g(e^*) = \{f \in F(G); e$ belongs to the boundary of $f\}$.

It is important that the dual graph is defined with respect to an embedding. In fact, a *planar* graph may have several non-isomorphic dual graphs associated with it, corresponding to its different planar embeddings.

If G is a topological planar graph then G^* is planar. There is a natural way to properly draw G^* to the plane: represent each dual vertex $f \in F(G)$ as a point on face f, and represent each dual edge e^* by a curve between the points representing its endvertices, which crosses exactly once the representation of e in G and is disjoint from the rest of the representations of both G and G^* (see Figure 2.14).

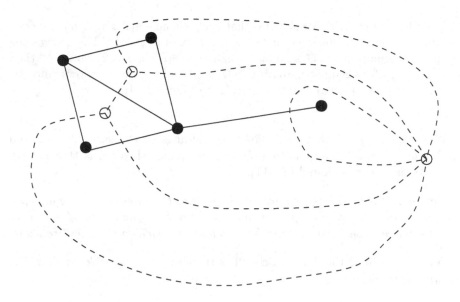

Figure 2.14. A planar graph and its dual

Let us recall that a set of edges is even if it induces even degrees at each vertex. We will say that a subset E' of edges of a plane graph G is *dual even* if $\{e^*; e \in E'\}$ is an even set of edges of G^*.

Observation 2.10.6. *The dual even subsets of edges of G are exactly the edge cuts of G.*

Whitney noticed that this observation characterizes planar graphs:

Definition 2.10.7. Let G, G^* be connected multigraphs and let f be a bijection between their edge-sets. We say that G^* is a *combinatorial dual* of G if, for each $F \subset E(G)$, F is the edge-set of a cycle of G if and only if $f(F)$ is a minimal (with respect to inclusion) edge cut in G^*.

The proof of the following Whitney's characterization of planarity can be found in [MT].

Theorem 2.10.8. *Let G be a 2-connected multigraph. Then G is planar if and only if it has a combinatorial dual. If G^* is a combinatorial dual of G, then G has an embedding in the plane such that G^* is isomorphic to the geometric dual of this embedding of G.*

Definition 2.10.9. A set of edges which is a cycle and bounds a face of a plane graph will be called a *facial cycle*.

Theorem 2.10.10. *Let G be a 2-connected plane graph. Then each face is bounded by a cycle of G and each edge belongs to the facial cycles of exactly two faces. If G is not a cycle, then G^* is also 2-connected.*

Proof. It follows from Theorem 2.1.6 that each 2-connected graph has an edge e so that $G - e$ is 2-connected, or G has a vertex of degree 2. This proves the first part. If a multigraph G with more than one vertex is not 2-connected then neither is any of its geometric duals G^*. The second part now follows since we can write (in a sloppy way) that $G^{**} = G$ (see Figure 2.14).

$\square$

When shall we say that two planar embeddings of the same 2-connected graph are the same? The following Jordan-Schönflies theorem is the starting point. Its proof can be found in [MT].

Theorem 2.10.11. *If f is a homeomorphism (i.e. bijection preserving the open sets, see Section 5.1) of a simple closed curve C in the plane onto a closed curve C' in the plane, then f can be extended to a homeomorphism of the entire plane.*

With the help of the Jordan-Schönflies theorem, it is possible to show the following (see[MT]):

Theorem 2.10.12. *Let G and G' be two 2-connected isomorphic plane graphs such that each facial cycle of G corresponds to a facial cycle of G' and the cycle bounding the outer face of G also bounds the outer face of G'. Then any homeomorphism of G and G' (extending an isomorphism of G onto G') can be extended to a homeomorphism of the entire plane.*

Hence we say that two planar embeddings of the same 2-connected graph are *equivalent* if their sets of facial cycles are the same. Next we characterize all non-equivalent planar embeddings of a 2-connected planar graph G. Let us call a *flip* a move which replaces a 2-connected component of a plane graph G by its rotation by 180 degrees around the axis determined by its 2-vertex cut. For an example see Figure 2.15.

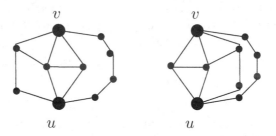

Figure 2.15. A flip

Theorem 2.10.13. *Let G be a 2-connected plane multigraph. Then any plane multigraph G' isomorphic to G is equivalent to an embedding obtained from G by a sequence of flips.*

Proof. The geometric duals of G and G' have the same cycle space since G is isomorphic to G'. Moreover it is not difficult to see that the flips on G correspond exactly to the twists on the corresponding geometric duals. Whitney's theorem (Theorem 2.3.6) therefore implies Theorem 2.10.13.

$\square$

In particular, we obtain that planar embeddings of 3-connected graphs are unique.

Corollary 2.10.14. *All planar embeddings of a 3-connected planar graph are equivalent.*

Next, we write down a combinatorial characterization of planar graphs, the *Kuratowski's theorem*. The notion of subdivision was used before, in Theorem 2.1.8.

Theorem 2.10.15. *A graph is planar if and only if it does not contain a subdivision of K_5 or a subdivision of $K_{3,3}$ as a subgraph.*

We leave as an exercise to show (using the Jordan Curve Theorem 2.10.1) that neither a subdivision of K_5 nor a subdivision of $K_{3,3}$ is planar. The other (difficult) part of the proof will follow directly from Lemma 2.10.16, Lemma 2.10.18 and Lemma 2.10.19 below.
We recall that G/e denotes the multigraph obtained from G by contraction of the edge e. In this part we are not interested in the loops or the multiple edges and so we delete them after each contraction. We hence stay in the class of graphs.

Lemma 2.10.16. *Every 3-connected graph $G = (V, E)$ with at least 5 vertices contains an edge e such that the graph G/e is 3-connected.*

Proof. Let G be a counterexample and let $xy = e \in E$. The graph G/e has a separating set consisting of two vertices. One of them must be the vertex obtained by contracting xy, since G is 3-connected. Hence G has a separating set $\{x, y, z\}$ and we choose x, y, z so that the largest component of $G - \{x, y, z\}$ is as large as possible. We denote it by H. Let H' be another component of $G - \{x, y, z\}$ and let u be a vertex of H' adjacent to z. Repeating the argument for uz, G has a separating set of the form z, u, v. The subgraph of G induced on $(V(H) \cup \{x, y\}) \setminus \{v\}$ is connected (otherwise v, z separate G), and thus it is contained in a connected component of $G - \{z, u, v\}$. This gives a contradiction to the maximality of H.

$\square$

We make use of two additional basic notions.

Definition 2.10.17. A plane graph is *straight line embedded* if each edge is represented as a straight line segment. If, in addition, each inner face is convex and the outer face is the complement of a convex set, then the graph is *convex embedded*.

Lemma 2.10.18. *Every 3-connected graph containing no subdivision of $K_{3,3}$ or K_5 as a subgraph has a convex embedding in the plane.*

Proof. We proceed by induction on the number of vertices of G. If G has at most 5 vertices then the lemma clearly holds, and we thus show the induction step. By Lemma 2.10.16, G has an edge $e = xy$ so that $G' = G/e$ is 3-connected. Let z denote the vertex obtained by identifying x and y. G' contains no subdivision of $K_{3,3}$ or K_5 (otherwise G does as well) and thus, by the induction assumption, G' has a convex embedding in the plane. The graph $G' - z$ is 2-connected and thus by Theorem 2.10.10, its face containing the point z is bounded by a cycle which we denote by C. Clearly, C is also a cycle in G. Let $x_1, \cdots, x_k$ be the neighbours of x, appearing along C in this order. If all the neighbours of y belong to a segment of C between some x_i, x_{i+1} then the convex embedding of G' may be easily modified to a convex embedding of G, otherwise G contains a subdivision of $K_{3,3}$ or K_5.

$\square$

Lemma 2.10.19. *Every graph with a fixed set of at least four vertices and a maximal set of edges with no subdivision of $K_{3,3}$ or K_5 as a subgraph is 3-connected.*

Proof. We again proceed by induction on the number of vertices of G. If G has at most 5 vertices then the lemma clearly holds, and we thus show the induction step. It is easy to verify that G must be 2-connected and if u, v separate G then $uv \in E$. If G is not 3-connected then it has two vertices u, v that separate it, and we may thus write $G = G_1 \cup G_2$ where G_1 and G_2 have precisely the vertices x, y and the edge xy in common. We observe that the addition of any edge creates a subdivision of $K_{3,3}$ or K_5 in each G_i and thus by the inducton assumption, each G_i is either K_3 or 3-connected. Hence each G_i has a convex embedding in the plane by Lemma 2.10.18. Let z_i be a vertex distinct from u, v but on the boundary of the same face of the convex embedding of G_i. By the assumption of the theorem, $G + z_1 z_2$ contains a subdivision K of $K_{3,3}$ or K_5. If all vertices of K of degree at least 3 are in G_1 (or in G_2), then we get a subdivision of $K_{3,3}$ or K_5 in G_1 (or in G_2), a contradiction. Since $K_{3,3}$ and K_5 are 3-connected, one of $V(G_1) \setminus V(G_2)$ or $V(G_1) \setminus V(G_2)$, say $V(G_2) \setminus V(G_1)$, contains precisely one vertex of K of degree at least 3. Since K_5 is 4-connected, K must be a subdivision of $K_{3,3}$. But this means that we can find a subdivision of $K_{3,3}$ in the graph obtained from G_1 by adding one new vertex and joining it to u, v, z_1. But all these neighbours are on the boundary of the same face of G_1, this graph is thus planar which gives a contradiction.

$\square$

As an immediate corollary we get Tutte's convex embedding theorem. Later in Section 5.4 we will include another proof.

Theorem 2.10.20. *Every 3-connected planar graph has a convex embedding in the plane.*

2.11   Tree-width and excluded minors

A graph H is a *minor* of a graph G if H can be obtained from a subgraph of G by *contracting* some connected subgraphs to single vertices. A *well-quasi-ordering*

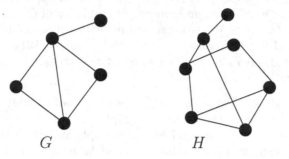

G H

Figure 2.16. G is a minor of H

of a set X is a reflexive and transitive relation $\preceq$ such that for every infinite sequence $x_1, x_2, \cdots$, of elements of X, there are indices $i < j$ such that $x_i \preceq x_j$. One of the highlights of modern combinatorics is the excluded minors theory of Robertson and Seymour. The main result is a (very complicated) proof of *Wagner's conjecture*: the class of all finite graphs with the minor relation is a well-quasi-ordering. As a straightforward corollary we get the following seminal theorem.

Theorem 2.11.1. *Let M be a minor-closed class of finite graphs. Then the set F of the graphs that are minimal graphs(with respect to the relation of being a minor) not in M is finite.*

Such a set of *minimal excluded minors* can be extremely large: there are some natural classes M for which this 'finiteness statement' may be formulated in *Finite Set Theory*, but cannot be proved or disproved there.
The main tool in obtaining the proof of the Wagner's conjecture is a structural theorem for the classes of graphs with an excluded minor. Another discovery of Robertson and Seymour is a polynomial algorithm for testing whether an input graph has a fixed minor.

Corollary 2.11.2. *Membership testing for any minor-closed class $\mathcal{C}$ of graphs admits a polynomial algorithm.*

Proof. As a consequence of Wagner's conjecture, there is a finite set S of graphs such that $G \in \mathcal{C}$ if and only if G does not have a minor from S. The corollary follows from the fact that testing minors is polynomial.

□

The theory only states the existence of the algorithm, the actual list of forbidden minors is usually hard to find. An important example of a minor-closed class of graphs is the class consisting of the finite graphs embeddable on a (fixed) surface S. As a corollary we get that a Kuratowski-type theorem holds for any S. A fundamental notion of the theory of Robertson and Seymour is *tree decomposition* and *tree width*.

Definition 2.11.3. Let $G = (V, E)$ be a graph and let $T = (W, F)$ be a tree where each vertex $w \in W$ has an induced subgraph G_w of G associated with it. We say that $T, (G_w : w \in W)$ is a *tree decomposition* if each edge of G belongs to some G_w, and for each vertex v of G, the set $\{w \in W; v \in G_w\}$ induces a connected subgraph T_v of T. The *width* of this tree decomposition is max $\{|V(G_i)| - 1; i = 1, \cdots, n\}$.

The *tree width* of a graph G, $tw(G)$, is the smallest width of a tree decomposition of G. Hence the connected graphs of tree width at most one are precisely the trees. See Figure 2.17 for an example of a graph of tree width 3. Tree decompositions are closely related to *chordal graphs*, i.e., graphs without

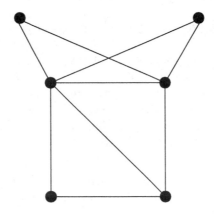

Figure 2.17. A graph of tree width 3

an induced cycle of length bigger than three. Here are some basic observations about chordal graphs, which are easy to prove.

Observation 2.11.4. *Let $G = (V, E)$ be a chordal graph. Each minimal cut induces a complete subgraph. There is a simplicial vertex, i.e. a vertex $v \in V$ such that $\{u \in V : \{u, v\} \in E\}$ induces a complete subgraph.*

Observation 2.11.5. *Let G be a chordal graph and let $T = (W, F), (G_w; w \in W)$ be its tree decomposition. Then G is the intersection graph of the subtrees $T_v, v \in V$ (see Definition 2.11.3),i.e., $uv \in E$ if and only if $T_u \cap T_v \neq \emptyset$.*

If $G = (V, E)$ is a graph then any chordal graph on V that contains G as a subgraph is called a *chordal completion* of G.

Corollary 2.11.6. *Let $G = (V, E)$ be a graph. Then $tw(G) = m - 1$, where m is the minimum, over all chordal completions G' of G, of the number of vertices in a maximum complete subgraph (clique) of G'.*

It follows from Corollary 2.11.6 that the class of graphs of tree width at most k is minor-closed. Another related observation is that a big planar square grid has a big tree width. It is much harder to show that the converse is also true. The proof of the following theorem can be found in [MT].

Theorem 2.11.7. *Let $r, m > 0$ be integers, and let G be a graph of tree-width at least $r^{4m^2(r+2)}$. Then G contains either K_m or the $r \times r$ planar square grid as a minor.*

Chapter 3

Trees and electrical networks

3.1 Minimum spanning tree and greedy algorithm

Given a connected graph $G = (V, E)$ and a weight function $w : E \to \mathbb{Q}$, the *minimum spanning tree problem* asks for a spanning tree $T = (V, F)$, $F \subset E$, for which $\sum_{e \in F} w(e)$ is minimal. A *greedy algorithm* (GA) is as follows: in the initial step we choose a cheapest edge, and in each subsequent step we choose one among the cheapest remaining edges with the restriction that the subgraph of G formed by the selected edges is acyclic.

The first algorithm to find the minimum spanning tree was proposed by Borůvka in the 1920's to plan the electrification of a region of Moravia. This algorithm works only if all the costs are different. Suppose the villages in an area are to be joined by electric wires. Each village (independently) starts bulding a connection, of course to the (unique, since all the weights are different) nearest village. It can (but need not) happen that both village x and village y build the connection xy. At the end of this stage some villages are connected but the whole system can be disconnected. Hence at the next stage the groups of the joined villages perform the same procedure as single villages at the previous step.

Theorem 3.1.1. *Both the greedy algorithm and Borůvka's algorithm produce a minimum spanning tree. If no two edges have the same cost then there is a unique minimum spanning tree.*

Proof. Let T_1 be a spanning tree constructed by the greedy algorithm and let T be a minimum spanning tree that has as many edges in common as possible with T_1. If $T \neq T_1$ then let e be the first edge added to T_1 that does not belong to T. Let P be the unique path in T between the endvertices of e. Since $P + e$ is a cycle, P must have an edge e' which doesn't belong to T_1. When e

was added to T_1 by the greedy algorithm, e' was also considered; we thus have $w(e) \leq w(e')$. Then, however, $T - e' + e$ is a minimum spanning tree with more edges in common with T_1, a contradiction. So $T = T_1$.

Suppose now that no two edges have the same cost. Let T_2 be the spanning tree constructed by Borůvka's method (Homework: why does Borůvka's method produce a spanning tree?). Let T be a minimum spanning tree. Again let e be the first edge not in T that was selected for T_2. Edge e was selected since it is the least costly connection from a subtree F to the rest of the vertices. The path between the endvertices of e in T has an edge e' joining a vertex of F to a vertex out of F and thus $w(e) < w(e')$. This is, however, impossible since $T - e' + e$ is also a spanning tree and it has a smaller total cost than T. We note that this also proves the uniqueness of a minimum spanning tree (assuming no two edges have the same cost.

$\square$

3.2   Tree isomorphism

It is an open problem to find an efficient way to recognize non-isomorphic graphs. We describe an efficient algorithm to distinguish non-isomorphic trees. First, however, let us spend some time for the general problem. There are two variants: we can either try to distinguish non-isomorphic graphs algorithmically, or we can try to find a natural function on graphs that separates non-isomorphic graphs. In the next short discussion on graph isomorphism we unfortunately need some concepts introduced later in the book. Still it seems that this is a natural place for such a discussion. Bollobás, Pebody and Riordan made the following conjecture (the intuitive notion of *almost all graphs* was made clear in Section 2.8, and the Tutte polynomial is introduced in Section 7.2).

Conjecture 3.2.1. *Almost all graphs are uniquely determined by their Tutte polynomial, i.e., for almost all graphs G there is no graph H so that G and H are not isomorphic and have the same Tutte polynomial.*

The function X_G and the q−dichromate are introduced in Section 7.5. Stanley asked whether

Conjecture 3.2.2. *The symmetric function generalization X_G of the chromatic polynomial distinguishes non-isomorphic trees.*

I propose

Conjecture 3.2.3. *All chordal graphs are distinguished by the q−dichromate.*

Replacing the Tutte polynomial by the q−dichromate should be a reasonable generalization in view of Theorem 7.5.10. Moreover, the isomorphism problem for general graphs can be reduced to the isomorphism problem in the class of chordal graphs: if $G = (V, E)$ is a graph then we can construct a chordal graph $ch(G) = (V', E')$ defined by $V' = V \cup E$ and $E' = (\binom{V}{2}) \cup \{ve : v \text{ incident}$

with e in G}. Clearly two graphs G and G' are isomorphic if and only if $ch(G)$ and $ch(G')$ are isomorphic. Conjecture 3.2.3 implies that each chordal graph can be determined from the system of its proper (w.r.t. the edge set or w.r.t. the vertex set) subgraphs. This concept is extensively studied. It is called edge reconstructability and vertex reconstructability. The edge reconstructability of chordal graphs was proved by Thatte. An important feature of chordal graphs is their tree structure, see Section 2.11. Each tree is obviously chordal. Conjecture 3.2.3 for trees is weaker than Conjecture 3.2.2. After all of these conjectures let us get back to the algorithmic testing of tree isomorphism.

Definition 3.2.4. A *rooted tree* is a tree with one distinguished vertex called the *root*. If the vertex u lies on the path from its neighbour v to the root then u is the *father* of v and v is a *son* of u. A *planted tree* (see Figure 3.1) is a rooted tree together with the fixed linear orderings of the sets of the sons of each vertex. This is naturally represented by a plane drawing of the tree, where the root is the lowest vertex and the orderings of the sons of each vertex are left-right orderings. Two rooted (planted) trees are isomorphic if they differ only by names of their vertices.

Let us denote the isomorphism of trees by $\cong$, the isomorphism of rooted trees by $\cong_1$ and the isomorphism of planted trees by $\cong_2$.
We first define, by Figure 3.1, a *code* $c(T)$ of a planted tree: Each leaf different from the root gets 01. If all sons $s_1, \cdots, s_k$ of a vertex v are assigned codes $c(s_1), \cdots, c(s_k)$, then v gets $c(v) = 0c(s_1) \cdots c(s_k)1$. Finally, the code $c(T)$ is defined to be the code of the root of T.

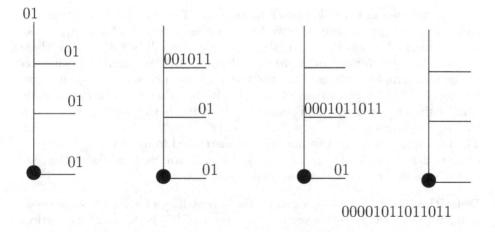

Figure 3.1. Planted tree and its code

Observation 3.2.5. *Two planted trees are isomorphic if and only if their codes are the same.*

Proof. We need to show that a code corresponds to a unique (up to renaming of the vertices) planted tree. We use the induction on the number of vertices. The beginning of the induction is straightforward. Each code has a form $c = (0A_1 \ldots A_k 1)$, where A_1 can be identified as the shortest segment which has the same number of 0's as 1's, and similarly we can detect $A_2, \ldots, A_k$. Each A_i is the code of a son s_i of the root, and s_i is to the left of s_j for $i < j$. Each A_i is also the code of a planted tree rooted at s_i and, by the induction assumption, we already know that A_i codes a unique planted tree. Hence also c codes a unique planted tree. $\square$

A simple decoding procedure is described in Figure 3.2.

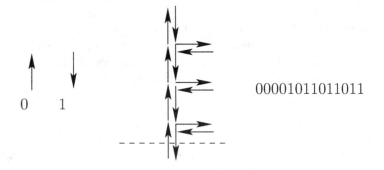

$$00001011011011$$

Figure 3.2. Decoding

Next we associate a code to each rooted tree. The leaves different from the root will again get 01, and once we have a father u whose all sons $s_1, \cdots, s_k$ already received codes $A_1, \ldots, A_k$, then the code of u will be $0A_1' \cdots A_k' 1$ where $A_1', \cdots, A_k'$ is the *lexicographic* ordering of $A_1 \cdots A_k$. We recall the definition of the lexicographic ordering from the first chapter: Let $a = (a_1, \cdots, a_n)$ and $b = (b_1, \cdots, b_m)$ be two strings of integers. We say that a is lexicographically smaller than b if a is an initial segment of b or, if j is the smallest index such that $a_j \neq b_j$, then $a_j < b_j$.

The rooted tree can again be uniquely reconstructed from its code. Finally, in order to define a code of a tree, we will find a *canonical root* and define the code of a tree as the code of the corresponding rooted tree.

Definition 3.2.6. Let G be a graph. The excentricity of a vertex is the maximum distance from the other vertices. The *center* $C(G)$ is the set of the vertices of G with minimum excentricity.

Observation 3.2.7. *For any tree T, $C(T)$ consists of one vertex or of two vertices connected by an edge.*

Proof. This follows from the observation that if we delete all the leaves of T, the center does not change. $\square$

The definition of the center clearly does not depend on the names of the vertices and so it is preserved under isomorphism. We can thus define the canonical root of a tree T as follows: if T has one-vertex center then let it be the root. Otherwise, if uv is the center, then $T-uv$ has exactly two components. Let us root them in u and v, make their codes, and root T in whichever u or v has a lexicographically smaller code.

3.3 Tree enumeration

We will calculate the number of the spanning trees of a graph. Let us start with *Cayley's formula*.

Theorem 3.3.1. *For each $n \geq 2$, the number of the spanning trees of the complete graph K_n equals n^{n-2}.*

Proof. Let us call a spanning tree with two marks (a circle and a square) placed on its (not necessarily distinct) vertices a *vertebrate*.
Since the number of the vertebrates equals n^2 times the number of the spanning trees and the number of the functions on $\{1, \cdots, n\}$ is n^n, the theorem follows from Lemma 3.3.2 below.

□

Lemma 3.3.2. *There exists a bijection between the set of all vertebrates and the set of all functions of $\{1, \cdots, n\}$ to itself.*

Proof. A vertebrate W has a unique path P between the marked vertices. Let us list the vertices of P first in the increasing order and then in the order as they appear on P when read from the circle to the square. This defines a permutation of the vertices of P. We write down its decomposition into directed cycles (see Figure 3.3). Next, there are subtrees of W 'hanging out' from the vertices of P; we direct all their edges towards the vertices of P. This finishes the construction of a directed graph $D(W)$. It follows that each vertex of $D(W)$ has exactly one directed edge leaving it, and so $D(W)$ defines a function f_W on $\{1, \cdots, n\}$. On the other hand, each function f on $\{1, \cdots, n\}$ uniquely determines a directed graph $D(W)$ for some W, and W may be uniquely obtained from $D(W)$. □

Next we express the number $T(G)$ of the spanning trees of a connected graph G as the determinant of a minor of the *Laplace matrix* $L(G) = (q_{uv})_{u,v \in V}$. The Laplace matrix is defined as follows: $q_{uu} = \deg_G(u)$, $q_{uv} = -1$ if $uv \in E$, and $q_{uv} = 0$ otherwise. Let $(L(G) : u,v)$ denote the minor of $L(G)$ obtained by removing row u and column v of $L(G)$.

Theorem 3.3.3. *For every connected graph G and $u \in V$, $T(G) = \det(L(G) : u,u)$.*

Proof. Let D be an arbitrary orientation of G and let us denote also by $I_D = (d_{ij})$ the incidence matrix of the orientation D, i.e. the $V \times E$ matrix defined by $d_{ue} = -1$ if vertex u is the tail of e in D, $d_{ue} = 1$ if u is the head of e in

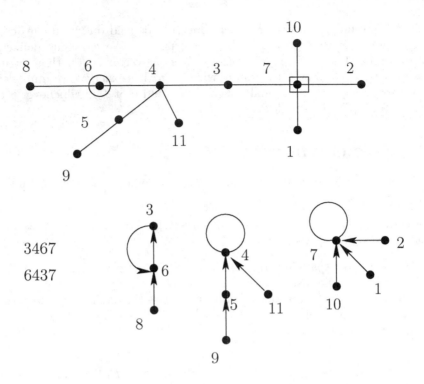

Figure 3.3. A vertebrate and the corresponding mapping

D, and $d_{ue} = 0$ otherwise. Let $(D : u)$ denote the matrix obtained from D by deleting row u. It is easy to check the following observation.

Observation 3.3.4. $DD^T = L(G)$ and $(D : u)(D : u)^T = (L(G) : u, u)$.

Here comes the key step of the proof.

Observation 3.3.5. *Let T be a spanning subgraph of G with $|V|-1$ edges. Let R be an orientation of T, and let u be a vertex of G. Then $\det(I_R : u) \in \{1, -1, 0\}$ and $\det(I_R : u) \neq 0$ if and only if T is a tree.*

Proof. If a vertex $v \neq u$ is isolated in T then $\det(I_R : u) = 0$ since it has a zero row. If there is $v \in V - u$ which has degree 1 in T then we can expand $\det(I_R : u)$ along the v-row, and consider a smaller matrix. If none of the two cases applies then each $v \neq u$ has degree 2 in T and u has degree 0. In this case $\det(I_R : u) = 0$

$\square$

To finish the proof of Theorem 3.3.3, we use the *Binet-Cauchy formula*. It says that for an arbitrary $n \times m$ matrix A, $\det(AA^T) = \sum_J \det(A_J)^2$, where the sum is over all n-element subsets $J \subset \{1, \cdots, m\}$ and A_J denotes the matrix obtained from A by deleting all columns whose indices do not belong to J. Hence, by the Binet-Cauchy formula,

$$\det(L(G) : u, u) = \det((I_D : u)(I_D : u)^T) = \sum_J \left(\det((I_D : u)_J) \right)^2.$$

By Observation 3.3.5 this is exactly the number of spanning trees of G.

$\square$

3.4 Electrical networks

The number of spanning trees of a graph was first studied in connection with calculations in *electrical networks*.

An *electrical network* is a multigraph $G = (V, E)$ where each edge e is assigned a real number $r(e)$ called *resistance*. The *conductance* of e is $c(e) = 1/r(e)$. Let $e = uv$. If there is a *potential difference* $p((uv))$ from u to v then the *electric current* $w(e)$ flows in e from u to v according to *Ohm's law* (OL):

$$w(uv) = \frac{p(uv)}{r(e)}.$$

Let's make this precise. We started with an undirected multigraph but the currents flow in some direction. If the potential difference in a directed edge (u, v) is $p(u, v)$ then the potential difference in the reversed edge (v, u) is $p(v, u) = -p(u, v)$. The functions p and w are governed by the laws of Kirchhoff:

Kirchhoff's potential (voltage) law (KPL) states that the potential differences sum to 0 around any cycle of G. KPL is equivalent to saying that there exist *absolute potentials* $V(u), u \in V$ so that for each edge $e = (uv)$, $p(uv) = V(u) - V(v)$. If the network is connected and the potential differences $p(e)$ are given, then we can choose arbitrarily the potential of one of the vertices (usually, it is chosen to be 0) and this determines the other vertex potentials.

Kirchhoff's current law (KCL) states that the total current entering a vertex v is equal to the total current leaving v:

$$E_v + \sum_{(x,v)\in E} w(x, v) = L_v + \sum_{(v,x)\in E} w(v, x),$$

where E_v denotes the amount of current entering v from outside of G and L_v denotes the amount of current leaving v to outside of G.

If the only vertices where a non-zero current enters or leaves the network are s, t, then by KCL the total current from s to t is $E_s = L_t$. If the potential difference from s to t is p, then by Ohm's law $r = p/E_s$ is the *total resistance* of the network between s and t.

A *solution* of the electrical network is the current function w. There are several tools in practical calculations: for a series connection the resistances add, for a parallel connection the conductances add, and two vertices with the same absolute potentials get *shorted* (identified); see Figure 3.4.

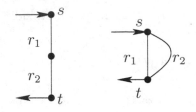

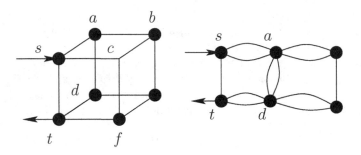

Figure 3.4. Resistors connected in series and in parallel; by symmetry (all resistors are assumed to be the same) $V(a) = V(c)$ and $V(d) = V(f)$ so a is identified with c and d is identified with f

The *star-triangle transformation* appears for the first time in this book: Let v be a vertex of degree three connected to vertices a, b, c (the *star*), and let no current be allowed to enter or leave the network at v. Then we can replace the star by the *triangle* as shown in Figure 3.5, where the resistances are calculated. The corresponding networks have precisely the same currents leaving vertices a, b, c.

A theorem of Kirchhoff states that any network has a solution. But first let us observe that there is always *at most one solution*.

Lemma 3.4.1. *Each network has at most one solution.*

Proof. For simplicity we consider only the case when all resistances are the same. If there were two different currents, i.e., functions on the set of the directed edges satisfying KPL, KCL and OL, then their difference is a non-zero current satisfying KPL, KCL and OL, in the network which no current enters or leaves. In a circulation, if a non-zero (say positive) current flows in some edge, it is pushed further and since the network is finite, there will be a directed cycle where a positive current flows. That contradicts KPL. □

Here comes the theorem of Kirchhoff.

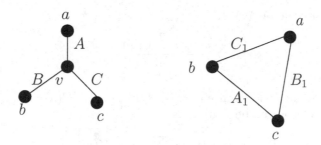

Figure 3.5. $A_1 = S/A, B_1 = S/B, C_1 = S/C, S = AB + BC + CA$

Theorem 3.4.2. *Let (uv) be a directed edge of a connected network G where each edge has a unit resistance. Let us denote by $N(s, u, v, t)$ the number of spanning trees of G in which the unique path from s to t contains u and v, in this order. We recall (see Theorem 3.3.3) that $T(G)$ denotes the total number of spanning trees of G. Let w be a function on the set of the directed edges of G defined by*

$$w(uv) = \frac{N(s, u, v, t) - N(s, v, u, t)}{T(G)}.$$

Then w is a current of unit size from s to t satisfying KCL, KPL, OL.

Proof. It follows from its definition that w satisfies OL and KPL. If T is a spanning tree then let $w(T)$ be the current of size 1 along the unique path P from s to t in T, i.e. $w(T)(uv) = 1$ if uv appears in P in this order, $w(T)(uv) = -1$ if uv appears in P in the opposite order, and $w(T) = 0$ otherwise. We have

$$N(s, u, v, t) - N(s, v, u, t) = \sum_T w(T)(uv).$$

The function $\sum_T w(T)$ satisfies KCL since it is a sum of elementary functions which clearly satisfy KCL.   □

Let us describe two constructions of a current in a network G, with no current leaving or entering the network at vertices other than $s_1, \cdots, s_k$. By Lemma 3.4.1, the two constructions output the same current.

The approach assuming KCL and OL. Consider a flow $f(uv)$ with outlets (sources and sinks) $s_1, \ldots, s_k$. In order to turn the flow to a proper electric current with outlets $s_1, \ldots, s_k$ all we have to make sure is that KPL holds, i.e., that

$$\sum_{e \in C} f(e) r(e) = 0$$

for every oriented cycle C. The flow f satisfying this condition is the current.

The approach assuming KPL and OL. Consider absolute potentials $V(u)$ on the vertices. This gives a current with outlets $s_1, \ldots, s_k$ if and only if KCL holds, i.e., for each $u \in V - \{s_1, \ldots, s_k\}$,

$$\sum_{uw \in E} \frac{V(u) - V(w)}{r(uw)} = 0.$$

We recall that the conductance $c(uv) = 1/r(uv)$. Let $C(u) = \sum_{\{u,v\} \in E} c(uv)$. The KCL condition can be written as

$$C(u)V(u) = \sum_{uv \in E} c(uv)V(v).$$

If $c(uv) = 1$ everywhere then this condition says that V is a *discrete harmonic function* with boundary $s_1, \ldots, s_k$.

Definition 3.4.3. Let $G = (V, E)$ be a graph and $S \subset V$. A real function f on V is said to be a *discrete harmonic function (DHF)* on G with boundary S if

$$f(x) = \frac{1}{\deg(x)} \sum_{\{x,y\} \in E} f(y)$$

whenever $x \in V(G) - S$ and $\deg(x) \geq 1$.

Let us now introduce the notion of the *energy* in the network. We are given an edge uv with resistance $r(uv)$ and potential difference $p(uv) = V(u) - V(v)$, and hence with a current of size $w(uv) = p(uv)/r(uv) = (V(u) - V(v))/r(uv)$. The *energy* in uv is defined to be

$$w^2(uv)r(uv) = \frac{(V(u) - V(v))^2}{r(uv)} = (V(u) - V(v))w(uv).$$

The *total energy* in the network is the sum of the energies of the edges, i.e.

$$\sum_{\{uv\} \in E} w^2(uv)r(uv).$$

Let us formulate Dirichlet's principle:

Theorem 3.4.4. *Let G be an electrical network with resistances $r(e)$, outlets $s_1, \cdots, s_k$ and absolute potentials $V(s_i), i = 1, \ldots, k$, for the outlets. Then there are absolute potentials $V(u), u \in V - \{s_1, \ldots, s_k\}$, so that the energy*

$$\sum_{\{uv\} \in E} \frac{(V(u) - V(v))^2}{r(uv)}$$

is minimal. These absolute potentials give a proper electric current with outlets among $s_1, \cdots, s_k$.

Proof. The energy is a continuous function of the absolute potentials and it goes to infinity when the maximum of the absolute potentials goes to infinity. This implies that the infimum of the energy is attained for some $V_0(u), u \in V - \{s_1, \ldots, s_k\}$. For this solution we must have the derivative of the energy with respect to each $V(u), u \in V - \{s_1, \ldots, s_k\}$, equal to zero. Hence for each $u \in V - \{s_1, \ldots, s_k\}$,

$$\sum_{\{uw\} \in E} \frac{2(V(u) - V(w))}{r(uw)} = 0.$$

Hence the solution defines the current satisfying KCL.

$\square$

Thompson's principle reads as follows:

Theorem 3.4.5. *Let G be an electrical network with resistances $r(e)$ and outlets $s_1, \cdots, s_k$. Let $f(s_i) \in \mathbb{R}$, $i = 1, \ldots, k$ such that $\sum_i f(s_i) = 0$. We again consider the energy function*

$$\sum_{\{u,v\} \in E} f^2(uv) r(uv)$$

for flows $f(uv)$ in which a flow of size $f(s_i)$ (possibly negative) enters the network at outlet s_i. Then there is such a flow f minimizing the energy, and this flow satisfies KPL and so it is a proper electric current.

Proof. Similarly as in the proof of Theorem 3.4.4 we observe that a flow f_0 minimizing the energy exists. For each fixed cycle C let $f(\epsilon)$ be the flow obtained from f_0 by increasing it in each edge of C by ϵ. The energy as a function of ϵ has its minimum for $\epsilon = 0$, and there the derivative of the energy with respect to ϵ equals zero, and so KPL holds for the cycle C.

$\square$

The next theorem is Rayleigh's conservation of energy principle.

Theorem 3.4.6. *Let $D = (V, E)$ be a directed graph and let f be an s,t-flow, i.e., a function on the directed edges satisfying KCL at each vertex other than s, t. Let $f(s) = \sum_u f(su) - \sum_v f(vs)$ and let $V(u)$ be any function on the vertices. Then*

$$(V(s) - V(t))f(s) = \sum_{uv \in E} (V(u) - V(v))f(uv).$$

Proof.

$$\sum_{uv \in E} (V(u) - V(v))f(uv) = \sum_{u \in V} V(u)(\sum_{uv \in E} f(uv) - \sum_{vu \in E} f(vu)) =$$

$$V(s)f(s) - V(t)f(t) = (V(s) - V(t))f(s).$$

$\square$

Corollary 3.4.7. *The total energy in an electric current from s to t is equal to* $(V(s) - V(t))w(s)$, *where* $w(s)$ *is the value of the current.*

The *effective conductance* $C_{\text{eff}} = C_{\text{eff}}(s, t)$ of an electrical network from s to t is the value of the current from s to t if s, t are set at potential difference 1. The *effective resistance* $R_{\text{eff}} = 1/C_{\text{eff}}$ is the potential difference between s, t ensuring a current of size 1 from s to t.

Dirichlet's principle and Thompson's principle have the following consequences.

Corollary 3.4.8.

$$C_{eff}(s, t) = \inf_V \left(\sum_{\{uv\} \in E} \frac{(V(u) - V(v))^2}{r(uv)} : V(s) = 1, V(t) = 0 \right),$$

$$R_{eff}(s, t) = \inf_f \left(\sum_{\{uv\} \in E} f^2(uv) r(uv) : f(xy) \text{ is an s-t flow of size 1} \right).$$

Corollary 3.4.9. *If the resistance of an edge is increased (in particular if the edge is deleted), then the effective resistance (between two vertices) doesn't decrease.*

3.5   Random walks

We learned in the previous section that the function of absolute potentials is a discrete harmonic function (DHF). Another natural source of DHFs are random walks on graphs. A (discrete time) *Markov chain* on a finite or countable set V of *states* is a sequence of random variables $X_0, X_1, \cdots$ taking values in V such that for all $x_0, \cdots, x_{t+1} \in V$, the probability of $X_{t+1} = x_{t+1}$, conditioned on $X_0 = x_0, \cdots, X_t = x_t$, depends only on x_t and x_{t+1}. Markov chains on graphs are called *random walks*.

We consider directed multigraph $G = (V, E)$ with positive edge-weights. Let a_{uv} be the sum of the weights of the multiple edges directed from u to v, and let $A_u = \sum_v a_{uv}$. Let P be the $V \times V$ matrix defined by

$$P_{uv} = a_{uv}/A_u.$$

Thus P is a matrix with non-negative entries in which each row-sum is 1. We interpret the matrix P as the *transition probability matrix* of a random walk on G: P_{uv} describes the probability of going from u to v in one step. The matrix P together with the initial distribution X_0 determines the random walk by

$$X_t = X_0 P^t.$$

The entries of the matrix P^t are called *t-step transition probabilities*. A state i is *accessible* from state j if for some t, $P_{j,i}^t > 0$. A Markov chain is *irreducible* if each state is accessible from any other state. Let $P_e(i, \infty)$ be the probability (called *escape probability*) that, when starting at state i, we never return back to it. A state i is *transient* if $P_e(i, \infty) > 0$, and it is *recurrent* otherwise. A random walk is recurrent if all its states are recurrent. We futher denote by $h_{i,i}$ the expected time to return to state i when starting at state i. A recurrent state i is *positive recurrent* if $h_{i,i} < \infty$. Finally, a state i is *periodic* if there is $\Delta > 1$ such that

$$\Pr(X_{t+s} = i | X_t = i) = 0$$

unless s is divisible by Δ. A Markov chain is periodic if all its states are periodic.

Definition 3.5.1. An aperiodic, positive recurrent state is an *ergodic* state. A Markov chain is ergodic if all its states are ergodic.

A *stationary distribution* of a Markov chain is a distribution π such that $\pi = \pi P$, where P is the corresponding transition matrix. The following two fundamental theorems on Markov chains characterize chains that converge to stationary distributions.

Theorem 3.5.2. *Any irreducible aperiodic Markov chain with a finite set of states $V = \{0, 1, \cdots, n\}$ has the following properties:*

(1) The chain has a unique stationary distribution π.

(2) For all j and i, the limit $\lim_{t \to \infty} P_{j,i}^t$ exists and it is independent of j.

(3) $\pi(i) = \lim_{t \to \infty} P_{j,i}^t = 1/h_{i,i}$.

Theorem 3.5.3. *Any irreducible aperiodic Markov chain with V countable infinite belongs to one of the following two categories:*

(1) The chain is ergodic; for all j and i, the limit $\lim_{t \to \infty} P_{j,i}^t$ exists and it is independent of j, and the chain has a unique stationary distribution $\pi(i) = \lim_{t \to \infty} P_{j,i}^t = 1/h_{i,i}$.

(2) No state is positive recurrent; for all j and i, $\lim_{t \to \infty} P_{j,i}^t = 0$, and the chain has no stationary distribution.

Random walks provide another construction of a current in an electrical network. The electrical network is a weighted multigraph where the weight is given by the conductance; we will denote by c_{uv} the sum of the conductances of all the parallel edges with the endvertices u, v. Let P denote the corresponding symmetric transition probability matrix.

Theorem 3.5.4. *Let G be a connected electrical network and let $s \neq t$ be vertices of G. Let us consider the random walk whose transition probability matrix P is defined by the conductances $c(u, v)$. For a vertex u let*

$$V(u) = \Pr(\text{starting at } u, \text{ we get to } s \text{ before we get to } t).$$

Hence $V(s) = 1$ and $V(t) = 0$. Then $V(u)$ gives the absolute potential for the unit current from s to t.

Proof. If vertex u is different from s, t then we have

$$V(u) = \sum_v P_{uv} V(v) = \sum_v \frac{c(uv)}{C(u)V(v)},$$

and hence

$$C(u)V(u) = \sum_v c(uv)V(v).$$

The theorem now follows from the fact proven earlier that each electrical network has a unique solution (Lemma 3.4.1, Theorem 3.4.2). $\qquad\square$

Chapter 4

Matroids

Matroids provide a successful connection between graph theory, geometry and linear algebra. Some of the dualities we will discuss later are rooted in the theory of matroids. Moreover, matroids provide a basis for discrete optimization. Several important algorithms, for instance the greedy algorithm, belong to the matroid world. We make a notational agreement in this chapter: the graphs are allowed to have loops and multiple edges.

Definition 4.0.5. Let X be a finite set and $S \subset 2^X$. We say that $M = (X, S)$ is a matroid if the following conditions are satisfied:

(I1) $\emptyset \in S$,

(I2) $A \in S$ and $A' \subset A$ then $A' \in S$ (S is *hereditary*),

(I3) $U, V \in S$ and $|U| = |V| + 1$ then there is $x \in U - V$ so that $V \cup \{x\} \in S$ (S satisfies an *exchange axiom*).

Example 4.0.6. Let X be the set of all columns of a matrix over a field and let S consist of all the subsets of X that are linearly independent. Then (X, S) is a matroid (called *vectorial or linear matroid*).

Definition 4.0.7. Let $M = (X, S)$ be a matroid. The elements of S are called *independent sets* of M. The maximal elements of S (w.r.t. inclusion) are called *bases*. Let $A \subset X$. The *rank* of A, $r(A)$, is defined by $r(A) = max\{|A'|; A' \subset A, A' \in S\}$. The *closure* of A, $\sigma(A)$, equals $\{x; r(A \cup \{x\}) = r(A)\}$. If $A = \sigma(A)$ then A is *closed*.

By repeated use of (I3) in Definition 4.0.5 we get

Corollary 4.0.8. *If $U, V \in S$ and $|U| > |V|$ then there is $Z \subset U - V$, $|Z| = |U - V|$ and $V \cup Z \in S$. All bases have the same cardinality.*

Theorem 4.0.9. *A non-empty collection $\mathcal{B}$ of subsets of X is the set of all bases of a matroid on X if and only if the following condition is satisfied.*

(B1) If $B_1, B_2 \in \mathcal{B}$ and $x \in B_1 - B_2$ then there is $y \in B_2 - B_1$ such that $B_1 - \{x\} \cup \{y\} \in \mathcal{B}$.

Proof. Property (B1) is true for matroids: we apply (I3) to $B_1 - \{x\}, B_2$. To show the other implication we need to prove that each hereditary system satisfying (B1) satisfies (I3) too. First we observe that (B1) implies that no element of $\mathcal{B}$ is a strict subset of another one, and by repeated application of (B1) we observe that in fact all the elements of $\mathcal{B}$ have the same size. To show (I3) let B_U, B_V be bases containing U, V from (I3) and such that their symmetric difference is as small as possible. If $(B_V \cap (U - V)) \neq \emptyset$ then any element from there may be added to V and (I3) holds. We show that $(B_V \cap (U - V)) = \emptyset$ leads to a contradiction with the choice of B_U, B_V: If $x \in B_V - B_U - V$ then (B1) produces a pair of bases with smaller symmetric difference. Hence $B_V - B_U - V$ is empty. But then necessarily $|B_V| < |B_U|$, a contradiction.   □

Theorem 4.0.10. *A collection S of subsets of X is the set of all independent sets of a matroid on X if and only if (I1), (I2) and the following condition are satisfied.*

(I3') If A is any subset of X then all the maximal (w.r.t. inclusion) subsets Y of A with $Y \in S$ have the same cardinality.

Proof. Property (I3') is clerly equivalent to (I3).   □

Theorem 4.0.11. *An integer function r on 2^X is a rank function of a matroid on X if and only if the following conditions are satisfied.*

(R1) $r(\emptyset) = 0$,

(R2) $r(Y) \leq r(Y \cup \{y\}) \leq r(Y) + 1$,

(R3) If $r(Y \cup \{y\}) = r(Y \cup \{z\}) = r(Y)$ then $r(Y) = r(Y \cup \{y, z\})$.

Proof. Clearly (R1),(R2) hold for matroids. To show (R3) let B be a maximal independent subset of Y. If $r(Y) < r(Y \cup \{y, z\})$ then B is not maximal independent in $Y \cup \{y, z\}$, but any enlargement leads to a contradiction.

To show the other direction we say that A is independent if $r(A) = |A|$. Obviously the set of the independent sets satisfies (I1). If A is independent and $B \subset A$ then $r(B) = |B|$ since otherwise, by (R2), $r(A) \leq |B - A| + r(B) < |A|$. Hence (I2) holds. If (I3) does not hold for U, V then by repeated application of (R3) we get that $r(V \cup (U - V)) = r(V)$, but this set contains U, a contradiction.   □

Theorem 4.0.12. *An integer function on 2^X is a rank function of a matroid on X if and only if the following conditions are satisfied.*

(R1') $0 \leq r(Y) \leq |Y|$,

(R2') $Z \subset Y$ implies $r(Z) \leq r(Y)$,

(R3') $r(Y \cup Z) + r(Y \cap Z) \leq r(Y) + r(Z)$. This property is called submodularity.

Proof. Clearly (R1') and (R2') hold for matroids. To show (R3') let B be a maximal independent set in $Y \cap Z$ and let B_Y, B_Z be maximal independent in Y, Z containing B. We have $r(Y \cap Z) = |B_Y \cap B_Z|$ and clearly $r(B_Y \cup B_Z) \leq |Y \cup Z|$. Hence (R3') follows. On the other hand, (R1),(R2) and (R3) follow easily from (R1'), (R2') and (R3').

$\square$

Theorem 4.0.13. *The closure $\sigma(A)$ is the smallest (w.r.t. inclusion) closed set containing A.*

Proof. First observe that $\sigma(A)$ is closed, since $r(\sigma(A) \cup \{x\}) = r(\sigma(A))$ implies $r(A \cup \{x\}) \leq r(\sigma(A) \cup \{x\}) = r(\sigma(A)) = r(A)$. To show the second part let $A \subset C$, C closed and $x \in (\sigma(A) - C)$. Hence $r(C \cup \{x\}) > r(C)$ and this implies $r(A \cup \{x\}) > r(A)$. (exercise: why?) This contradicts $x \in \sigma(A)$.

$\square$

Theorem 4.0.14. *A function $\sigma : 2^X \to 2^X$ is the closure operator of a matroid on X if and only if the following conditions are satisfied.*

(S1) $Y \subset \sigma(Y)$,

(S2) $Z \subset Y$ then $\sigma(Z) \subset \sigma(Y)$,

(S3) $\sigma(\sigma(Y)) = \sigma(Y)$,

(S4) if $y \notin \sigma(Y)$ but $y \in \sigma(Y \cup \{z\})$ then $z \in \sigma(Y \cup \{y\})$. This property is called the Steinitz-MacLane exchange axiom.

We say that two matroids are isomorphic if they differ only in the names of their groundset elements.

4.1 Examples of matroids

We already know vectorial matroids. A matroid is *representable* if it is isomorphic to a vectorial matroid.

Let $G = (V, E)$ be a graph and let $M(G) = (E, S)$ where $S = \{F \subset E; F \text{ forest}\}$. Then $M(G)$ is a matroid, called the *cycle matroid* of G. Its rank function is $r(F) = |V| - c(F)$, where we recall that $c(F)$ denotes the number of connected components of the spanning subgraph (V, F). The matroids isomorphic to cycle matroids of graphs are called *graphic matroids*.

Let $G = (V, E)$ be a graph. The *matching matroid* of G is the pair (V, S) where $A \in S$ if and only if A may be covered by a matching of G. This is a matroid since the basis axiom corresponds to the exchange along an alternating path of two maximum matchings of G.

A matroid is *simple* if $r(A) = |A|$ whenever $|A| < 3$. Simple matroids of rank 3 have a natural representation that we now describe. Each matroid is determined by its rank function and so each simple matroid M of rank 3 is determined by the set $L(M) = \{A \subset X; |A| > 2, r(A) = 2, A \text{ closed }\}$; if $|A| > 2$ then $r(A) = 2$ if and only if A is a subset of an element of $L(M)$.

Lemma 4.1.1. *If $A, B \in L(M)$ then $|A \cap B| \leq 1$.*

Proof. We assume for a contradiction $\{x, z\} \subset A \cap B$, $a \in A - B$ and $b \in B - A$. Then both a, b belong to $\sigma(\{x, z\})$ and hence, by Theorem 4.0.13, both a, b belong to any closed set containing $\{x, z\}$: a contradiction.

$\square$

A set $C \subset 2^X$ is a *configuration* on X if each element of C has at least 3 elements and any pair of elements of C have at most one element of X in common.

Theorem 4.1.2. *Each configuration is the set $L(M)$ of a simple matroid of rank 3 on X.*

Proof. Given C, for each $A \subset X$ define $r(A) = |A|$ if $|A| \leq 2$, and if $|A| > 2$ then $r(A) = 2$ if and only if A is a subset of an element of C; $r(A) = 3$ otherwise. We show that r is a rank function of a matroid. Note that $(R1, (R2)$ are obviously satisfied. We show $(R3)$: If $r(Y \cup \{y\}) = r(Y \cup \{z\}) = r(Y)$ then $|Y| \geq 2$ and both $Y \cup \{y\}$, $Y \cup \{z\}$ are subsets of an element of C. They are in fact subsets of the same element of C since their intersection has size 2. Hence $r(Y) = r(Y \cup \{y, z\})$.

$\square$

Hence we can represent simple matroids of rank 3 by a system of 'lines' in the plane corresponding to the elements of $L(M)$. The most famous picture of matroid theory, the *Fano matroid F_7*, is depicted in Figure 4.1. The Fano matroid is the vectorial matroid, over $GF(2)$, of the matrix whose columns are all non-zero vectors of $GF(2)^3$.

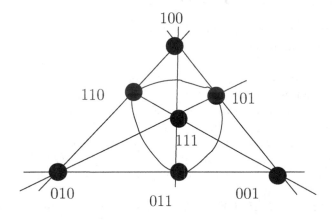

Figure 4.1. Fano matroid F_7

4.2 Greedy algorithm

Let (X, S) be a set system and w a weight function on $X = \{1, 2, \cdots, n\}$. In a *discrete optimization problem* we may want to find $J \in S$ such that $\sum_{i \in J} w_i$ is maximized. We encountered the *greedy algorithm* (GA) in Section 3.1. There, it was shown that GA correctly solves the minimum spanning tree problem (in order to turn the minimum spanning tree problem into a maximization problem we change the sign of each weight). Let us first define the greedy algorithm in a more general way, as an algorithm for the general optimisation problem that works as follows:

- Order the elements of X so that $w_1 \geq w_2 \geq \cdots \geq w_n$.

- $J := \emptyset$.

- For $i = 1, \cdots, n$ do: if $J \cup \{i\} \in S$ and $w_i \geq 0$ then $J := J \cup \{i\}$.

The next theorem shows that applicability of GA characterizes matroids.

Theorem 4.2.1. *Let (X, S) be a hereditary non-empty set system. Then the greedy algorithm solves the discrete optimization problem correctly for any weight function w on X if and only if (X, S) is a matroid.*

Proof. If a hereditary system is not a matroid then it does not satisfy (I3') and it is not difficult to construct a weight function w for which the greedy algorithm does not work. Let us prove the opposite implication: Let m be maximal such that $w_m \geq 0$. Let z' be the characteristic vector of a set produced by the greedy algorithm and let z be the characteristic vector of any other set of S. Let $T_i = \{1, \cdots, i\}$, $i = 1, \cdots, m$. We notice that for each i

$$z'(T_i) = \sum_{j \leq i} z'_j \geq \sum_{j \leq i} z_j = z(T_i),$$

since $J \cap T_i$ is a maximal subset of T_i which belongs to S (by the definition of GA). We have

$$wz \leq \sum_{i=1}^{m} w_i z_i = \sum_{i=1}^{m} w_i(z(T_i) - z(T_{i-1})) =$$

$$\sum_{i=1}^{m-1} (w_i - w_{i+1})z(T_i) + w_m z(T_m) \leq \sum_{i=1}^{m-1} (w_i - w_{i+1})z'(T_i) + w_m z'(T_m) = wz'.$$

$\square$

The only property we used in the proof is that $z \geq 0$ and $z(T_i) \leq z'(T_i) = r(T_i)$. GA thus solves also the following problem:

maximize $\sum_{i \in X} w_i z_i$

$z(A) = \sum_{i \in A} z_i \leq r(A)$, $A \subset X$;

$z_i \geq 0$, $i \in X$.
The problems that may be described in this form are called *linear programs*, and the part of optimization which studies linear programs is called *linear programming*.

Corollary 4.2.2. *Edmonds Matroid Polytope theorem: For any matroid, the convex hull of the characteristic vectors of the independent sets is equal to* $\mathcal{P} = \{z \geq 0; \text{ for each } A \subset X, z(A) \leq r(A)\}$.

Proof. (sketch) The convex hull is clearly a subset of $\mathcal{P}$. By the Minkowski-Weyl theorem introduced in the beginning of the book we have that $\mathcal{P}$, a bounded intersection of finitely many half-spaces, is a *polytope*, i.e. a convex hull of its *vertices*. Each vertex c of $\mathcal{P}$ is characterized by the existence of a half-space $\{z; wz \leq b\}$ which intersects $\mathcal{P}$ exactly in $\{c\}$. Since GA solves any problem $\max\{wz; z \in \mathcal{P}\}$, each non-empty intersection of $\mathcal{P}$ with a half-space necessarily contains the incidence vector of an independent set. In particular, each vertex of $\mathcal{P}$ is the incidence vector of an independent set, and the theorem follows. □

Finally we remark that the greedy algorithm is polynomial time if there is a polynomial algorithm to answer the questions 'Is J independent ?'. It is usual for matroids to be given, for algorithmic purposes, by such an independence-testing oracle.

4.3   Circuits

Definition 4.3.1. A *circuit* in a matroid is a minimal (w.r.t. inclusion) non-empty dependent set.

The circuits of graphic matroids are the cycles of the underlying graphs.

Theorem 4.3.2. *A non-empty set $\mathcal{C}$ is the set of the circuits of a matroid if and only if the following conditions are satisfied.*

(C1) If $C_1 \neq C_2$ are circuits then C_1 is not a subset of C_2,

(C2) If $C_1 \neq C_2$ are circuits and $z \in C_1 \cap C_2$ then $(C_1 \cup C_2) - z$ contains a circuit.

Proof. First we show that a matroid satisfies the above properties. The first is obvious. For the second we have $r(C_1 \cup C_2) \leq r(C_1) + r(C_2) - r(C_1 \cap C_2) = |C_1| + |C_2| - |C_1 \cap C_2| - 2 = |C_1 \cup C_2| - 2$. Hence $(C_1 \cup C_2) - z$ must be dependent. On the other hand, we define S to be the set of all subsets which do not contain an element of $\mathcal{C}$ and show that (X, S) is a matroid. Axioms (I1) and (I2) are obvious and we show (I3'): let $A \subset X$ and for a contradiction let J_1, J_2 be maximal subsets of A that belong to S and $|J_1| < |J_2|$, and let $|J_1 \cap J_2|$ be as large as possible. Let $x \in J_1 - J_2$ and C the unique circuit of $J_2 \cup x$. Necessarily there is $f \in C - J_1$ and $J_3 = (J_2 \cup x) - f$ belongs to S by the uniqueness of C. Then $|J_3 \cap J_1| < |J_2 \cap J_1|$, a contradiction. □

Corollary 4.3.3. *If A is independent, then $A \cup \{x\}$ contains at most one circuit.*

Proposition 4.3.4. *Let $A \subset X$ and $x \notin A$. Then $x \in \sigma(A)$ if and only if there is a circuit C with $x \in C \subset A \cup \{x\}$.*

Proof. If $x \in \sigma(A)$ and B is maximal independent in A, then $B \cup x$ is dependent and hence contains a circuit. On the other hand, let D be a maximal independent set in A containing $C - x$. Then D is also maximal independent in $A \cup x$ and hence $x \in \sigma(A)$. $\qquad\square$

4.4 Basic operations

Definition 4.4.1. A *k-truncation* of M is a matroid M' on X such that A is independent in M' if and only if $|A| \leq k$ and A is independent in M.

Each truncation of a matroid is a matroid.

Definition 4.4.2. Let M_1, M_2 be matroids and $X_1 \cap X_2 = \emptyset$. $M_1 + M_2$ (direct sum of M_1, M_2) is the matroid on $X_1 \cup X_2$ such that A is independent if and only if $A \cap X_1$ is independent in M_1 and $A \cap X_2$ is independent in M_2.

Definition 4.4.3. Let X be a disjoint union of $X_i, i = 1, \cdots, n$ and let $S_i = \{A \subset X_i; |A| \leq 1\}$. Then $\sum_i (X_i, S_i)$ is called a *partition matroid*.

It follows immediately from the definition that $M \setminus U = (X \setminus U, S|_{X \setminus U})$ is a matroid. This operation is called *deletion of U*.

Definition 4.4.4. Let $T \subset X$ and let J be a maximal independent subset of $T' = X \setminus T$. M/T' (contraction of T') is a matroid on T defined so that A is independent if and only if $A \cup J$ is independent in M.

Theorem 4.4.5. *M/T' is a matroid and its rank function r' satisfies $r'(A) = r(A \cup T) - r(T)$. Hence M/T' does not depend on the choice of J.*

Proof. Obviously M/T' satisfies (I1) and (I2). Let $A \subset T$ and let J' be a maximal subset of A that is independent in M/T'. Observe that $J \cup J'$ is maximal independent in $A \cup T'$, by the choices of J, J'. $\qquad\square$

4.5 Duality

Definition 4.5.1. Let $M = (X, S)$ be a matroid. Its *dual matroid* is $M^* = (X, S^*)$ such that $I \in S^*$ if and only if $r(X \setminus I) = r(X)$ (r is the rank of M).

Proposition 4.5.2. *M^* is a matroid and its rank function r^* satisfies $r^*(A) = |A| - r(X) + r(X \setminus A)$.*

Proof. Again the only nontrivial property is (I3'). Let $A \subset X$ and let J be a maximal subset of A which belongs to S^*. Let B be a maximal independent (in M) subset of $X \setminus A$ and let B' be a basis of M containing B and $B' \subset X \setminus J$. If there is $x \in (A \setminus J) \setminus B'$ then J was not maximal (a contradiction). Hence $A \setminus J \subset B'$ and the formula for r^* follows. $\qquad\square$

The objects (bases, circuits, closed sets) of M^* are called dual objects or coobjects, e.g., dual bases or cobases. Let us note some simple facts: $M^{**} = M$. The dual bases are exactly complements of the bases. The cocircuits are minimal (w.r.t. inclusion) sets intersecting each basis. The cocircuits are exactly complements of hyperplanes. A *hyperplane* of M is a closed set whose rank is one less than $r(X)$).

Proposition 4.5.3. *Let G be a graph. Then the cocircuits of the graphic matroid $M(G)$ are exactly the minimal edge cuts.*

Proof. Note that edge cuts are exactly the sets of edges intersecting each basis of $M(G)$. □

Corollary 4.5.4. *Let G be a planar graph and G^* its geometric dual. Then $M(G^*) = M(G)^*$.*

Definition 4.5.5. M is called a *minor* of N if M is obtained from N by some finite sequence of deletions and contractions.

Let G be a graph. A minor of G is a graph obtained from G by deletions and contractions of edges. Observe the following: H is a minor of G if and only if $M(H)$ is a minor of $M(G)$.
The following series of propositions are proved by comparing the rank functions (we recall that the rank function uniquely determines the matroid).

Proposition 4.5.6. *We have*

(1) $(M/T)^* = M^* \setminus T$,

(2) $(M \setminus T)^* = M^*/T$,

(3) M is a minor of N if and only if M^* is a minor of N^*,

(4) M is a minor of N if and only if M may be obtained from N by a deletion (contraction) followed by a contraction (deletion).

A matroid M is called *cographic* if it is isomorphic to $M^*(G)$ for some graph G. It is also called a *cocycle matroid* of G. For example, it is not difficult to observe that $U_4^2 = (\{1, 2, 3, 4\}, \{\emptyset, 1, 2, 3, 4, 12, 13, 14, 23, 24, 34\})$ is not cographic. Next we recall Kuratowski's theorem (Theorem 2.10.15): G is planar if and only if G has no minor isomorphic to K_5 or $K_{3,3}$.

Proposition 4.5.7. $M(K_5)$ *and* $M(K_{3,3})$ *are not cographic.*

Proof. Assume $M(K_{3,3}) = M^*(G)$. Then $|E(G)| = 9$, G is a simple graph because no pair of edges separates $K_{3,3}$, and each edge cut of G contains at least 4 edges. Hence each degree of G is at least 4 and we get $4|V(G)| \leq 18$: a contradiction because G is simple. For K_5 one can use the fact that such a graph G has no circuit of length 3. □

Next comes a restatement of a classical theorem of Whitney about planar graphs.

Theorem 4.5.8. *G is planar if and only if its cycle matroid is cographic.*

Proof. By Corollary 4.5.4, if G is planar then $M(G) = M^*(G^*)$. To show the other direction, using the Kuratowski theorem, it suffices to observe that a minor of a cographic matroid is cographic (by dualizing the statement that a minor of a graphic matroid is graphic), and use Proposition 4.5.7. □

Here is an equivalent formulation: a matroid M is both graphic and cographic if and only if M is the cycle matroid of a planar graph.

4.6 Representable matroids

A matroid is called *binary* if it is representable over the 2-element field $GF(2)$. It is called *regular* if it is representable over an arbitrary field. Let A be a matrix representing matroid M and let A' be obtained from A by operations of adding a row to another row. Then again A' represents M. A representation of a matroid M is called *standard* w.r.t. a basis B if it has the form $I|A$, where I is the identity matrix of $r(M)$ rows whose columns are indexed by the elements of B. Since the elementary row operations do not change the matroid, we get that each representable matroid has a standard representation w.r.t. an arbitrary basis.

Theorem 4.6.1. *Let $I|A$ be a standard representation of M. Then $A^T|I$ is a representation of M^*.*

Corollary 4.6.2. *If M is representable over a field $\mathbb{F}$ and N is a minor of M then both M^* and N are representable over $\mathbb{F}$.*

Proof. Deletion clearly corresponds to deletion of the corresponding column in a representation. For contraction we use Theorem 4.6.1 and the duality between contraction and deletion.

□

Clearly, U_2^4 is not binary. Hence binary matroids do not have U_2^4 as a minor. Next we list some seminal results of Tutte, characterizing classes of matroids by forbidden minors.

Theorem 4.6.3. *M is binary if and only if M does not have U_2^4 as a minor. M is regular if and only if M is binary and does not have F_7 or F_7^* as a minor. M is graphic if and only if M is regular and does not have $M(K_5)^*$ or $M(K_{3,3})^*$ as a minor.*

We recall that F_7 denotes the Fano matroid. It is easy to observe that the graphic matroids are regular: Let $D = (V, E)$ be an arbitrary orientation of G and let I_D be the incidence matrix of D (see Section 2.3). Then I_D represents $M(G)$ over an arbitrary field, since a set of columns is linearly dependent if and only if its index set contains a cycle of G.

4.7   Matroid intersection

Given two matroids on the same set X, the matroid intersection problem is to
find a common independent set of maximum cardinality. Let us mention two
special cases: maximum matching in bipartite graphs (here the two matroids
are partition matroids), and maximum branching in a digraph (branching is a
forest in which each node has in-degree at most one); here one of the matroids
is the corresponding graphic matroid and the second one is a partition matroid
of the set-system of sets of the incoming edges at each vertex.

Theorem 4.7.1. *For two matroids (X, S_1) and (X, S_2), the maximum $|J|$ such
that $J \in S_1 \cap S_2$ equals the minimum of $r_1(A) + r_2(X \setminus A)$, over all $A \subset X$.*

Proof. If $J \in S_1 \cap S_2$ then for each $A \subset X$, $J \cap A \in S_1$ and $J \cap (X \setminus A) \in S_2$. Hence
$|J| \leq r_1(A) + r_2(X \setminus A)$. The second part is proved by induction on $|X|$. Let k
equal the minimum of $r_1(A) + r_2(X \setminus A)$ and let x be such that $\{x\} \in S_1 \cap S_2$.
Note: if there is no such x then $k = 0$, and if we take $A = \{x; r_1(\{x\}) = 0\}$, we
are done. Let $X' = X - x$. If the minimum over $A \subset X'$ of $r_1(A) + r_2(X \setminus A)$
also equals k then we are done by the induction assumption. Let S'_i denote S_i
contracted on $X \setminus x$. If the minimum over $A \subset X'$ of $r'_1(A) + r'_2(X \setminus A)$ is at
least $k - 1$ then the induction gives a common independent set of S'_1, S'_2 of size
$k - 1$ and adding x gives the desired common independent set of S_1, S_2. If none
of these happens, then there are $A, B \subset X'$ so that

$$r_1(A) + r_2(X' \setminus A) \leq k - 1$$

and

$$r_1(B \cup \{x\}) - 1 + r_2((X' \setminus B) \cup \{x\}) - 1 \leq k - 2.$$

Adding and applying submodularity we get

$$r_1(A \cup B \cup \{x\}) + r_1(A \cap B) + r_2(X \setminus (A \cap B)) + r_2(X \setminus (A \cup B \cup \{x\})) \leq 2k - 1.$$

It follows that the sum of the middle two terms or the sum of the outer two
terms is at most $k - 1$, a contradiction.   □

A polynomial time algorithm exists provided the rank can be found in polynomial time, even for the weighted case, but we do not include this here.

4.8   Matroid union and min-max theorems

The matroid union is closely related to the matroid intersection, as we will see.

Theorem 4.8.1. *Let $M' = (X', S')$ be a matroid and f an arbitrary function
from X' to X. Let $S = \{f(I); I \in S'\}$. Then (X, S) is a matroid with rank
function*

$$r(U) = min_{T \subset U}\{|U - T| + r'(f^{-1}(T))\}.$$

Proof. It suffices to show the formula for the rank function since obviously S is non-empty and hereditary. The formula follows from Theorem 4.7.1 since $r(U)$ is equal to the maximum size of a common independent set of M' and the partition matroid (X', W) induced by the family $(f^{-1}(s); s \in U)$. $\square$

Definition 4.8.2. If $M_i = (X_i, S_i), i = 1, \cdots, k$ are matroids and $X = \cup X_i$ then their union is defined as $(X, \{I_1 \cup I_2 \cdots \cup I_k; I_i \in S_i\})$.

Corollary 4.8.3. *Matroid union (partitioning) theorem: The union of matroids is again a matroid, with its rank function given by*

$$r(U) = min_{T \subset U}\{|U - T| + r_1(T \cap X_1) + \cdots + r_k(T \cap X_k)\}.$$

Proof. We first make X_i mutually disjoint and then use Theorem 4.8.1. $\square$

Example 4.8.4. Let $G = (V, W, E)$ be a bipartite graph. For each $u \in V$ define a matroid M_u on the set of neighbours of u so that a set is independent if and only if its cardinality is at most one. Then the union of $M_u, u \in V$ is called the *transversal matroid*.

Corollary 4.8.5. *The maximum size of a union of k independent sets of a matroid M is*

$$min_{T \subset X}\{|X \setminus T| + kr(U)\}.$$

Corollary 4.8.6. X *can be covered by k independent sets if and only if for each* $U \subset X$,

$$kr(U) \geq |U|.$$

Proof. X can be covered by k independent sets if and only if there is a union of k independent sets of size $|X|$.

$\square$

Corollary 4.8.7. *There are k disjoint bases if and only if for each* $U \subset X$,

$$k(r(X) - r(U)) \leq |X - U|.$$

Proof. There are k disjoint bases if and only if the maximum size of the union of k independent sets is $kr(X)$.

$\square$

Corollary 4.8.8. *A finite subset X of a vector space can be covered by k linearly independent sets if and only if for each* $U \subset X$,

$$k.r(U) \geq |U|.$$

These are some examples of *min-max theorems*, the pillars of discrete optimization.

Chapter 5

Geometric representations of graphs

The geometric representations of graphs provide some of the most useful decorations of abstract graphs. This chapter introduces basic concepts.

5.1 Topological spaces

A lot of the results described in this book depend on a representation of a graph on a 2−dimensional surface. In this preparatory section we review a few concepts from general topology. We begin by recalling the definition of a topological space, which is a mathematical structure capturing the notion of continuity.

Definition 5.1.1. A topological space is a pair $(X, \mathcal{O})$ where X is a (typically infinite) ground set and $\mathcal{O} \subset 2^X$ is a set system whose members are called the *open sets*; the open sets satisfy: $\emptyset \in \mathcal{O}$, $X \in \mathcal{O}$, the intersection of finitely many open sets is an open set, and so is the union of an arbitrary collection of open sets.

For example, in the standard topology of $\mathbb{R}^d$, a set $U \subset \mathbb{R}^d$ is open if and only if for every point $x \in U$ there exists $\varepsilon > 0$ such that the ε-ball $\{x : ||x|| \leq \varepsilon\}$ around x is contained in U. The same definition applies for any *metric space*. Even though the standard topology is induced by the Euclidean metric, the notion of 'being the same' for topological spaces is different than for the corresponding metric spaces. 'Being the same' in topological spaces is called a *homeomorphism*. Homeomorphism of topological spaces is defined in the same way as other isomorphisms we encountered so far. Two spaces $(X_1, \mathcal{O}_1)$ and $(X_2, \mathcal{O}_2)$ are *homeomorphic* if one is obtained from the other by renaming the elements of the ground-set. An example of the homeomorphic topological spaces is the real line $\mathbb{R}$ and the open interval $(0, 1)$.

Let $(X, \mathcal{O})$ be a topological space. Every $Y \subset X$ defines a *subspace*, namely

$(Y, \{U \cap Y; U \in \mathcal{O}\})$. A topological space $(X, \mathcal{O})$ is *Hausdorff* if for every two distinct points $x, y \in X$ there are disjoint open sets U, V with $x \in U$ and $y \in V$. A set $Y \subset X$ is *closed* if $X \setminus Y$ is open. The *closure* of $Y \subset X$ is the intersection of all closed sets containing Y, and the *boundary* of Y is the intersection of the closure of Y and the closure of $X \setminus Y$. If $(X_1, \mathcal{O}_1)$ and $(X_2, \mathcal{O}_2)$ are topological spaces, a mapping $f : X_1 \to X_2$ is called *continuous* if preimages of open sets are open.

Let X be a topological space. A *curve* (or an *arc*) in X is the image of a continuous function $f : [0, 1] \to X$. The curve *connects* its endpoints $f(0)$ and $f(1)$. A curve is *closed* if $f(0) = f(1)$. A topological space is *(arcwise) connected* if any two elements are connected by an arc in X. We will not consider other notions of connectivity of topological spaces.

Finally, a space $(X, \mathcal{O})$ is called *compact* if for every collection of open sets $\mathcal{U}$ with $\bigcup \mathcal{U} = X$, there exists a finite subcollection $\mathcal{U}_0 \subset \mathcal{U}$ with $\bigcup \mathcal{U}_0 = X$. In a compact metric space, every infinite sequence has a convergent subsequence. A space $X \subset \mathbb{R}^d$ is compact if and only if X is a closed and bounded set.

We want to study 2-dimensional surfaces, so here comes their definition.

Definition 5.1.2. A *surface* is a connected compact Hausdorff topological space S which is locally homeomorphic to an open disc in the plane, i.e., each point of S has an open neighborhood homeomorphic to the open unit disc in $\mathbb{R}^2$.

The surfaces may be constructed from regular polygons by glueing together their sides. But, before saying this more precisely, we introduce the simplices and the simplicial complexes.

Definition 5.1.3. A simplex σ is the convex hull of a finite affinely independent set $A \subset \mathbb{R}^d$. The points of A are called *vertices* of σ. The *dimension* of σ is $\dim \sigma = |A| - 1$. A k-dimensional simplex is called a *k-simplex*.

The convex hull of an arbitrary subset of vertices of a simplex σ is a *face* of σ. Thus every face is itself a simplex.

Definition 5.1.4. A non-empty family Δ of simplices is a *simplicial complex* if the following two conditions hold:

(1) Each face of any simplex $\sigma \in \Delta$ is also a simplex of Δ,

(2) The intersection $\sigma_1 \cap \sigma_2$ of any two simplices of Δ is a face of both σ_1 and σ_2.

The union of all simplices in a simplicial complex Δ is the *polyhedron* of Δ and is denoted by $P(\Delta)$. The *dimension* of a simplicial complex is the largest dimension of a simplex of the complex. The *vertex-set* of Δ, denoted by $V(\Delta)$, is the union of the sets of vertices of all simplices of Δ. The simplicial complex that consists only of the empty simplex is defined to have dimension -1. Zero-dimensional simplicial complexes are just configurations of points, and one-dimensional simplicial complexes correspond to graphs, represented geometrically with straight edges that do not cross.

Observation 5.1.5. *The set of all faces of a simplex is a simplicial complex.*

If f_i denotes the number of faces of dimension i in a simplicial complex Δ, then the *Euler characteristic* of Δ is the alternating sum

$$\sum_i (-1)^i f_i.$$

Next we show that a simplicial complex may be regarded as a purely combinatorial object. A collection K of subsets of set V is *hereditary* if $A \in K$ and $B \subset A$ implies $B \in K$. Every hereditary system (V, K) of subsets of V will be called an *abstract simplicial complex*. Its dimension $\dim(K) = \max\{|F| - 1; F \in K\}$. Each simplicial complex determines an abstract simplicial complex: $V = V(\Delta)$, and K is the set of the vertex-sets of all the simplices of Δ. On the other hand, it is easy to see that any abstract simplicial complex (V, K) with V finite has a geometric realization. Let $n = |V| - 1$ and let us identify V with the vertex-set of an n-simplex σ^n. We define a subcomplex Δ of σ^n by $\Delta = \{\text{conv}(F); F \in K\}$. This is a simplicial complex, and its associated abstract simplicial complex is (V, K). In fact, a much sharper result is true.

Theorem 5.1.6. *If Δ_1 and Δ_2 are two geometric realizations of an abstract simplicial complex then their polyhedra $P(\Delta_1)$ and $P(\Delta_2)$ are homeomorphic. Every finite abstract d-dimensional simplicial complex K has a geometric realization in $\mathbb{R}^{2d+1}$.*

The proof may be found in [MJ]. An important example of a simplicial complex is formed by the partially ordered sets. We recall that a *partially ordered set*, or *poset* for short, is a pair $(P, \preceq)$, where P is a set and $\preceq$ is a binary relation on P that is reflexive ($x \preceq x$), transitive ($x \preceq y$ and $y \preceq z$ imply $x \preceq z$), and weakly antisymmetric ($x \preceq y$ and $y \preceq x$ imply $x = y$). Now, if $(P, \preceq)$ is a poset then its *order complex* is the abstract simplicial complex whose vertices are the elements of P and whose simplices are all the chains (i.e. all the linearly ordered subsets, of the form ($\{x_1, \cdots, x_k\}$, $x_1 \prec x_2 \prec \cdots \prec x_k$) of P. We remark that this construction also naturally associates a poset to each simplicial complex.

Let X be a topological space. A simplicial complex Δ such that $P(\Delta)$ is homeomorphic to X, is called a *triangulation* of X. For example, a natural triangulation of the sphere S^{d-1} is the boundary of a d-simplex, that is, the subcomplex consisting of all of its proper faces. Other triangulations may be obtained from convex polytopes in $\mathbb{R}^d$.

Now we are ready to describe surfaces. We recall Definition 5.1.2: a *surface* is a connected compact Hausdorff topological space S which is locally homeomorphic to an open disc in the plane, i.e., each point of S has an open neighborhood homeomorphic to the open unit disc in $\mathbb{R}^2$. The next theorem expresses that we can get all surfaces by glueing together triangles. Its proof can be found in [MT].

Theorem 5.1.7. *Every surface has a finite triangulation of dimension* 2.

Let us consider two disjoint triangles T_1, T_2 with all sides equal, in a 2-simplex F of a triangulation of a surface S. We can make a new surface S' from S by deleting from F the interiors of T_1, T_2 and identifying T_1 with T_2 such that the clockwise orientations (in $F \subset \mathbb{R}^2$) around T_1 and T_2 disagree; see (a) of Figure 5.1 where the arrows indicate how the sides are identified. We say that S' is obtained from S by *adding a handle*. There is another possibility of identifying T_1 with T_2, as indicated in (b) of Figure 5.1. We say that the resulting surface S'' was obtained from S by adding a *twisted handle*. Finally let T be a quadrangle (with equilateral sides) in F. We let S''' denote the surface obtained from S by deleting the interior of T and identifying diametrically opposite points of the quadrangle as shown in (c) of Figure 5.1. We say that S''' is obtained from S by adding a *crosscap*. Let us consider now all surfaces

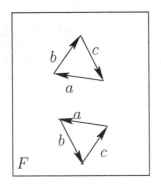
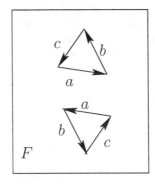
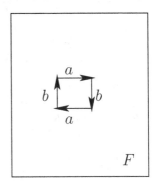

Figure 5.1. Adding a handle, a twisted handle and a crosscap

obtained from the sphere S_0 (which we can think of here as a tetrahedron) by adding handles, twisted handles and crosscaps. If we add h handles to S_0, we obtain S_h, the *orientable surface of genus* h. If we add h crosscaps to S_0 we get N_h, the nonorientable surface of genus h. The surface S_1 is the *torus* (the doughnut surface), N_1 is the *projective plane*, and N_2 is the *Klein bottle*. The Klein bottle cannot be realized as a subset of $\mathbb{R}^3$.

It is not difficult to observe that the location and the order of adding handles and crosscaps is not important: the resulting surface is always the same, up to homeomorphism. Adding a twisted handle amounts to the same, up to homeomorphism, as adding two crosscaps. Moreover, if we have already added a crosscap, then adding a handle amounts to the same, up to homeomorphism, as adding a twisted handle. In particular, if S is the surface obtained from the sphere by adding h handles, t twisted handles and c crosscaps then $S = S_h$ provided $t = c = 0$ and $S = N_{2h+2t+c}$ otherwise.

Now we are ready to state the *classification theorem for surfaces*. The proof can be found in [MT].

Theorem 5.1.8. *Every surface is homeomorphic to precisely one of the surfaces S_h or N_k.*

Next we extend the concept of a triangulation of dimension 2 to embeddings of graphs. Let X be a topological space. Analogously as in the Euclidean space, we say that a curve in X is the image of a continuous function $f : [0,1] \to X$. The curve is *simple* if f is one-to-one, and it *connects* its endpoints $f(0)$ and $f(1)$. A curve is *closed* if $f(0) = f(1)$. A topological space is *(arcwise) connected* if any two elements are connected by an arc in X. A set $C \subset X$ *separates* X if $X - C$ is not connected. A *face* of $C \subset X$ is a maximal connected component of $X - C$.

A graph G is *embedded* in a topological space X if the vertices of G are distinct elements of X and every edge of G is a simple arc connecting its two endvertices in X and such that its interior is disjoint from other edges or vertices. It is easy to see that every graph has an embedding in $\mathbb{R}^3$. A graph embedded in a topological space X is also called a *topological graph*. If G is a topological graph then we denote by $F(G)$ the set of its faces.

In this book, we embed graphs almost exclusively on surfaces. The notion which generalizes a triangulation is that of a map.

Definition 5.1.9. A *map* is a topological graph embedded on an orientable surface so that each face is homeomorphic to an open disc in the plane.

The next observation is Euler's formula, for the proof look again at [MT].

Lemma 5.1.10. *Let S_h be an orientable surface of genus h and let G be a map in S_h with v vertices, e edges and p faces. Then $v - e + p = 2 - 2h$.*

The genus of a map is usually defined as the genus h of the surface S_h where the map exists. Lemma 5.1.10 justifies us to define the *Euler characteristic* of the surface S_h in analogy to simplicial complexes by $\chi(S_h) = 2 - 2h$.

What is a map? The definition of a map is presented above, but Edmonds realized that maps may also be defined purely combinatorially. The proof of the next theorem can be found in [MT].

Theorem 5.1.11. *There is a natural bijection between maps and the connected graphs decorated with fixed cyclic orderings of the incident edges of each vertex.*

Physicists sometimes prefere *fatgraphs* to maps. This term corresponds to a helpful graphic representation of a graph (not necessarily connected), in which the vertices are made into discs (islands) and connected by fattened edges (bridges) prescribed by the cyclic orderings of the incident edges of each vertex. This defines a two-dimensional orientable surface with boundary. We usually denote a fatgraph and also the corresponding surface with boundary by F. Each component of the boundary of F is called a *face* of F. Each face is an embedded circle. We will denote by $G(F)$ the underlying graph of F. In Figure 5.2, the edges of $G(F)$ are denoted by dotted lines. We denote by $e(F), v(F), p(F), c(F), g(F)$ the number of edges, vertices, faces, connected components, and the genus of F. Let us rewrite the Euler formula for fatgraphs:

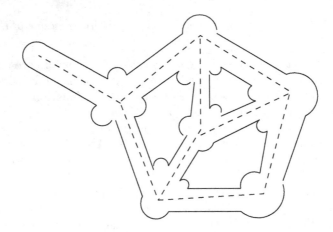

Figure 5.2. A fatgraph

Lemma 5.1.12. $v(F) - e(F) + p(F) = 2(c(F) - g(F))$.

An important concept is that of the *dual graph* G^* of a topological graph G. It may be defined for general topological graphs in exactly the same way as for the topological planar graphs, see Section 2.10.

5.2   Planar curves: Gauß codes

A closed curve C on a surface that crosses itself in a finite set I of points through which it passes exactly twice, gives rise to a word consisting of two copies of each element of I, as in Figure 5.3. Such a word is the *Gauß code* of the curve C, denoted by $Gauß(C)$. Gauß noticed some properties of these sequences for planar curves (see[G]). For $i \in I$ let $\Delta(i)$ denote the subset of I formed by the elements that appear exactly once between the two occurences of i. We say that i, j are *interlaced* if $j \in \Delta(i)$. The *interlaced graph* is the graph with vertex-set I and edge-set consisting of the interlaced pairs of letters.

Theorem 5.2.1. *A word w where each element of a finite set I appears exactly twice is a Gauß code of a planar curve if and only if the following conditions are satisfied:*

(1) *Any letter of w has an even number of interlaced letters.*

(2) *Any two noninterlaced letters have an even number of interlaced letters in common.*

(3) *The pairs of interlaced letters which have an even number of interlaced letters in common form an edge-cut in the interlaced graph.*

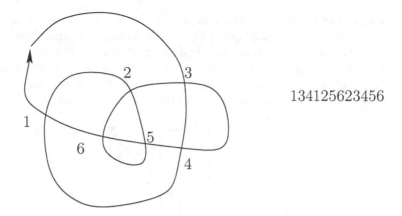

134125623456

Figure 5.3. Gauß code: an example

We present a construction of Rosenstiehl (see [RP] and for an independent proof see [FOM]) leading to a proof of Theorem 5.2.1. Given a fatgraph F, we construct its *medial graph* $M(F)$ as follows (see Figure 5.4): for each bridge, we 'cross' (or twist) its boundaries. The vertices of $M(F)$ are these crossings, and the edges are the connections left from the fatgraph F. The parts of former bridges become faces of $M(F)$. We call them the *discs* and color them black. The remaining faces of $M(F)$ will be colored white. This gives the *checkerboard* coloring of the faces of $M(F)$. Each vertex of $M(F)$ has degree 4. If $M(F)$ is

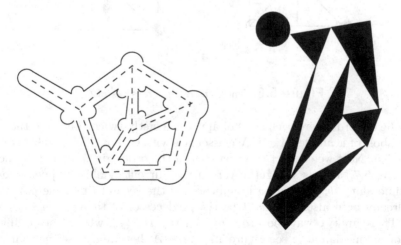

Figure 5.4. A fat graph and its medial graph

connected then it has an Euler tour and if we are even more lucky, there will be an Euler tour C which crosses itself at each vertex of $M(F)$. Such a tour C is a closed curve on the surface where F lives, and it gives rise to a Gauß code; the crossings of C are the vertices of $M(F)$ which correspond to the edges of the underlying topological graph of F: we will denote this topological graph also by F. Let F^* denote the geometric dual of F; let us draw it properly along with F. What we have got so far looks locally at each edge of F as in Figure 5.5, where the edges of F are solid, the edges of F^* are dotted, and the edges of $M(F)$ are bold. We naturally associate a sign to each crossing of C, as described in

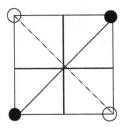

Figure 5.5. A patch around an edge of F

Figure 5.6. Theorem 5.2.1 may be proved by reversing this construction:

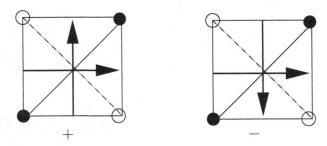

Figure 5.6. The signs of the crossings of C

Let w be a word where each symbol appears exactly twice. Let I be the set of the symbols of w and let $A \subset I$. We associate with each $e \in I$ a patch $P(e)$ as in Figure 5.5. Next we connect these patches in the order given by w, as indicated by Figure 5.7. This is straightforward if the current symbol e appears for the first time along w; if e currently appears for the second time, the patch $P(e)$ has already been inserted. Let f be the predecessor of the current symbol (e) in w. There are two ways how to connect $P(e)$ to $P(f)$, which lead to different signs of the median at e: see Figure 5.7. If $e \in A$ then we choose the connection which results in the positive sign and if $e \notin A$ then we choose the connection which results in the negative sign.

Observation 5.2.2. *In this way we construct a topological graph $F = F(w, A)$ on a surface (not necessarily orientable) along with its dual F^* and a checker-board coloring of its faces. An example is given in Figure 5.7. The edges of*

$F(w, A)$ are indexed by the set I of the symbols of w. Moreover the medial graph $M(F)$ admits an Euler tour $C = C(w, A)$ which crosses itself in each vertex of $M(F)$ and such that A is the set of the crossings of C of the positive sign. Finally, w equals the Gauß code of C.

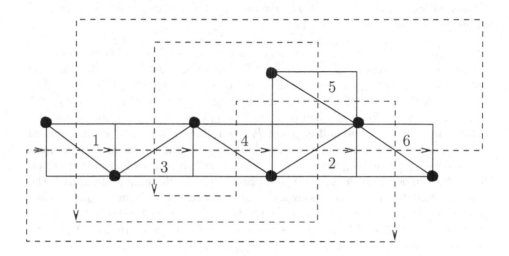

Figure 5.7. $w = 134261253456$, $A = \{1, 2, 6\}$

The Euler tour C of $M(F)$ from Observation 5.2.2 induces a walk P_C on the edges of F so that in P_C, each edge of F is traversed exactly twice. Analogously, C induces a walk P_C^* on the edges of F^* so that in P_C^*, each edge of F^* is traversed exactly twice. It is not difficult to verify

Observation 5.2.3. *The walk P_C traverses an edge e twice in the same direction if and only if the walk P_C^* traverses the edge e^* in the opposite directions if and only if $e \in A$.*

Let $I(e)$ denote the (incidence vector of the) subset of I of the symbols interlaced with e in w. We will use a symbol e to denote the characteristic vector of set $\{e\}$ as well. Next we introduce two functions on I.

Definition 5.2.4. We let $\alpha(e) = I(e) + e$ if $e \in A$ and $\alpha(e) = I(e)$ if $e \notin A$.
    We let $\beta(e) = I(e)$ if $e \in A$ and $\beta(e) = I(e) + e$ if $e \notin A$.

Observation 5.2.3 leads to the following

Corollary 5.2.5. *For each $e \in I$, $\alpha(e)$ belongs to the cycle space of F over $GF(2)$ and $\beta(e)$ belongs to the cycle space of F^* over $GF(2)$.*

The vertices of $M(F)$ partition each edge e of F into two *half-edges* e_1, e_2. The following straightforward characterization of $\alpha(e)$ will be useful.

Observation 5.2.6. *Let us start walking along P_C with edge e of F, and let e_1 be the initial half-edge of our walk. Let us stop at the moment when we are going to use the same half-edge e_1 for the second time (in either direction). Then $\alpha(e)$ is the set of edges of F whose both half-edges we traversed exactly once.*

Observation 5.2.7. $\{\alpha(e); e \in I\}$ *generates the cycle space of F and* $\{\beta(e); e \in I\}$ *generates the cycle space of F^*.*

Proof. We need to show that each vector z from the orthogonal complement of $\{\alpha(e); e \in I\}$ belongs to the cut space of F. We recall from Section 2.3 that the cut space is the orthogonal complement of the cycle space.

Given such z, let p be a function on the half-edges of F so that $p(e_1) + p(e_2) = 1$ modulo 2 if e_1, e_2 form an edge of z, and $p(e_1) + p(e_2) = 0$ modulo 2 for any other pair of consecutive half-edges of P_C. The function p may be constructed by choosing $p(h) = 0$ for an arbitrary half-edge h, and following the defining rules along P_C. Knowing that $z\alpha(e) = 0$ modulo 2, it follows from Observation 5.2.6 that we never get into a contradiction. The rules defining p along with the way we constructed F as a topological graph embedded on a surface imply that $p(h) = p(h')$ whenever h, h' share a vertex of F. Hence p induces a function $\pi : V(F) \to \{0,1\}$ so that z is the set of the edges uv of F such that $\pi(u) \neq \pi(v)$. Hence z belongs to the cut space of F.

$\square$

Corollary 5.2.8. *A word w where each element of a finite set I appears exactly twice is a Gauß code of a planar curve C if and only if there is $A \subset I$ such that* $\{\alpha(e); e \in I\}$ *is orthogonal to* $\{\beta(e); e \in I\}$.

Proof. A closed curve C in the plane with checkerboard coloring of the faces may be interpreted as an Euler tour of the median $M(F)$ of a planar map F. The set A corresponds to the sets of the vertices of C of the same sign. The cycle spaces of F and F^* are orthogonal since F is planar, and hence $\{\alpha(e); e \in I\}$ is orthogonal to $\{\beta(e); e \in I\}$ by Observation 5.2.7.

On the other hand, let w satisfy the conditions of Corollary 5.2.8. Let us denote by $\mathcal{K}$ the cycle space of $F = F(w, A)$, by $\mathcal{K}'$ the subspace of $\mathcal{K}$ generated by the faces of F, and by $\mathcal{K}''$ the cycle space of F^*. We have

$$\mathcal{K}' \subset \mathcal{K} =< \{\alpha(e); e \in I\} >\subset< \{\beta(e); e \in I\} >^*= [\mathcal{K}'']^* = K',$$

which implies that $\mathcal{K}' = \mathcal{K}$ and thus w is the Gauß code of a planar curve. $\square$

It is not difficult to observe that the conditions of Theorem 5.2.1 are necessary. Hence the following observation completes the proof of Theorem 5.2.1.

Observation 5.2.9. *The conditions of Theorem 5.2.1 hold if and only if we have that* $\{\alpha(e); e \in I\}$ *is orthogonal to* $\{\beta(e); e \in I\}$, *where α and β are defined w.r.t. set A which forms one of the vertex classes of the edge-cut of Condition (3).*

5.3 Planar curves: rotation

Function f is a *regular closed curve* if f is a continuously differentiable function from the closed unit interval $[0,1]$ into the plane $\mathbb{R}^2$ with $f(0) = f(1)$, and the derivative $f'(t)$ is non-zero everywhere. We consider regular closed curve as a *directed* curve. We recall that a closed curve is *simple* if it is one-to-one. It is customary that the points on a regular curve that are not one-to-one are called the *singularities*. A regular closed curve is *normal* if all its singularities form a finite number of simple crossings through which the curve passes exactly twice.

Definition 5.3.1. The *rotation* $\mathrm{rot}(C)$ of a regular closed curve C is the number of complete turns the tangent vector to the curve makes when passing once around the curve; the anti-clockwise turns are counted positive and the clockwise turns are counted negative.

The following theorem has been proved by Whitney.

Theorem 5.3.2. *Two regular closed curves in the plane can be continuously transformed into each other (without creating singularities) if and only if they have the same rotation number.*

A basic property of rotation is its additivity: if C is a concatenation of curves C_1 and C_2 and the tangent vectors are the same at the concatenation then $\mathrm{rot}(C) = \mathrm{rot}(C_1) + \mathrm{rot}(C_2)$. For normal closed curves there is a natural procedure of obtaining the rotation.

Lemma 5.3.3. *Let C be a normal closed curve in the plane. We can split it into disjoint cycles as indicated in Figure 5.8. Then $\mathrm{rot}(C)$ equals the number of anti-clockwise oriented cycles minus the number of clockwise oriented cycles from the split-decomposition of C.*

Proof. We smoothen C at each crossing. Imagine that we start at a crossing w and walk along C, until we return back to w for the first time. This splits C into C_1 and C_2 and clearly $\mathrm{rot}(C) = \mathrm{rot}(C_1) + \mathrm{rot}(C_2)$. The lemma follows e.g. by induction on the number of crossings, see Figure 5.8.

$\square$

Observation 5.3.4. *Let C be a normal closed curve and let c be the number of the crossings of C. Then $(-1)^{1+\mathrm{rot}(C)} = (-1)^c$.*

Proof. The curve C determines a 4-regular plane graph G whose vertices are the crossings. The cycles into which the curve decomposes as in Lemma 5.3.3 are the boundary cycles of the faces of G. The observation now follows from the Euler formula (Lemma 5.1.12).

$\square$

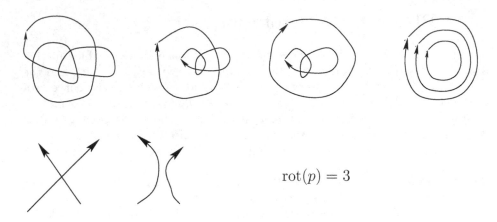

$$\mathrm{rot}(p) = 3$$

Figure 5.8. Rotation: an example

5.4   Convex embeddings

We first return to convex embeddings of Section 2.10 and show how they lead to the third natural construction of discrete harmonic functions. We recall that other sources of the discrete harmonic functions we discussed in the previous sections are absolute potentials in electrical networks, and random walks.

Let us recall Definition 2.10.17. A plane graph is *straight line embedded* if each edge is a straight line segment. If in addition each inner face is convex and the outer face is a complement of a convex set, then the graph is *convex embedded*.

Lemma 5.4.1. *Let G be a 2-connected planar graph. A straight line plane representation of G is convex if and only if the vertices of the outer face are embedded on a convex polygon in the compatible order, and each other vertex belongs to the convex closure of its neighbours.*

Proof. First let G be convex embedded and let v be an inner vertex. Since each face incident to v is convex, we can add straight lines between the neighbours of v to form a cycle on the neighbours with v inside. Hence v is in the convex closure of its neighbours. On the other hand, if an inner face is not convex and v_1, v_2, v_3 are its three consecutive vertices which define a concave angle at v_2, then v_2 belongs neither to the convex closure of its neighbours, nor to the outer face.

$\square$

Definition 5.4.2. Let $G = (V, E)$ be a planar 3-connected graph and let F be a face of G (we recall that by Corollary 2.10.14 the set of the faces does not depend on the planar embedding). A planar embedding f of its vertices obtained by fixing the vertices of F along a convex polygon in the compatible order, and for each $u \notin F$ letting

$$f(u) = \frac{1}{\deg(u)} \sum_{\{u,v\} \in E} f(v),$$

is called a *barycentric representation*.

Let us recall Tutte's theorem (Theorem 2.10.20): Every 3-connected planar graph has a convex embedding in the plane. In his proof, Tutte shows that each barycentric embedding of a 3-connected planar graph is a convex embedding. Let us further recall Definition 3.4.3 of a discrete harmonic function (DHF): A real function f on V is a DHF on G with boundary S if

$$f(u) = \frac{1}{\deg(u)} \sum_{\{u,v\} \in E} f(v)$$

whenever $u \in V - S$. This means that each coordinate of a barycentric embedding is DHF with boundary F. Hence Lemma 3.4.1 and Dirichlet's principle (Theorem 3.4.4) imply:

Corollary 5.4.3. *Let G be a planar 3-connected graph. Assume we fix an embedding of the vertices of one face of G along a convex polygon, in the compatible order. Then the barycentric representation of G which extends this fixed partial embedding is uniquely determined.*

Next we assume that the edges of G are made of ideal springs with unit Hooke constant, i.e., it takes h units of force to stretch them to length h. We nail the vertices of a face F to the vertices of a regular convex polygon. Then we let the system find its equilibrium. Let us denote by f the function that gives the planar position of each vertex. The energy of f is given by

$$\mathcal{E}(f) = \sum_{uv \in E} \|f(u) - f(v)\|^2,$$

where $\|.\|$ is the Euclidean norm. Dirichlet's principle (Theorem 3.4.4) applied for unit resistances and the connection with DHF shows the following:

Theorem 5.4.4. *The barycentric representation coincides with the unique f minimizing $\mathcal{E}(f)$.*

We have seen three sources of discrete harmonic functions (Definition 3.4.3) so far: Dirichlet's principle (Theorem 3.4.4), Theorem 3.5.4 from random walks, and ideal springs. These different constructions yield really the same object, by the uniqueness of harmonic extension.

Of course we can apply Dirichlet's principle for systems of ideal springs with general Hooke constants, and in general Euclidean space $\mathbb{R}^n$. Linial, Lovász and Wigderson noticed a beautiful application to graph connectivity, which we now describe.

Let $G = (V, E)$ be a graph and let U, W be two subsets of V. We denote by $p(U, W)$ the maximum number of vertex disjoint paths from U to W (disjointness includes the endvertices). We say that $X \subset \mathbb{R}^d$ is in *general position* if rank$(Y) = d + 1$ for each $Y \subset X$ of $d + 1$ elements (see the first chapter for the definition of the rank).

Definition 5.4.5. Let G be a graph and $X \subset V$. A *convex X-embedding* of G is any mapping $f : V \to \mathbb{R}^{|X|-1}$ such that each vertex of $V - X$ belongs to the convex hull of its neighbours. Further such an embedding is said to be in general position if the set $f(V)$ is in general position.

Theorem 5.4.6. *A graph G is k-connected $(1 < k < |V|)$ if and only if for every $X \subset V$, $|X| = k$, G has a convex X-embedding in general position.*

We first prove two lemmas.

Lemma 5.4.7. *Let G be a graph, $X \subset V$, f a convex X-embedding of G and $\emptyset \neq U \subset V$. Then $f(U)$ has at most $p(U, X)$ affinely independent points.*

Proof. By Menger's theorem, there is an $S \subset V$, $|S| = p(U, X)$, such that $G - S$ contains no path between X and U. Let W be the union of the connected components of $G - S - X$. Let $w \in W$. Since $w \notin X$, $f(w)$ lies in the convex hull of its neighbours and hence in the convex hull of $f(W \cup S - w)$. Hence $f(w)$ cannot be an extreme point of the convex hull of $f(W \cup S)$. It follows that $f(W)$ is a subset of the convex hull of $f(S)$. Hence

$$\text{rank}(f(U)) \leq \text{rank}(f(W \cup S)) = \text{rank}(f(S)) \leq |S| = p(U, X).$$

$\square$

Lemma 5.4.8. *Let G be a graph and $X \subset V$. Then G has a convex X-embedding f such that for every $\emptyset \neq U \subset V$, $f(U)$ has $p(U, X)$ affinely independent points.*

Proof. Let $|X| = k$. We start constructing the embedding f by assigning the elements of X to the vertices of a k-simplex in $\mathbb{R}^{k-1}$. Further we assign to each edge uv a positive *elasticity coefficient* $c_{uv} > 0$ and let $c_v = \sum_{uv \in E} c_{uv}$. Using Dirichlet's principle 3.4.4, we can uniquely extend f to f_c on V so that for each $u \notin X$,

$$f_c(u) = 1/c_u \sum_{uv \in E} c_{uv} f_c(v).$$

Hence each f_c is a convex X-embedding of G.
Let $U \subset V$ and $p(U, X) \geq 1$, otherwise the statement of the lemma clearly holds. We can also assume w.l.o.g. that $|U| = p(U, X)$; let $U = \{u_1, \cdots, u_m\}$. The defining linear equations of a discrete harmonic function imply that the coefficients of each $f(u_i)$ are rational functions of the coefficient-vector c. The set $f(U)$ is affinely dependent if and only if $\{f(u_1) - f(u_m), \ldots, f(u_{m-1}) - f(u_m)\}$ is linearly dependent if and only if $D(c) = 0$, where D is the sum of

the determinants of all the maximal square submatrices of the matrix with rows $f(u_1 - f(u_m)), \ldots, f(u_{m-1}) - f(u_m)$.

Finally we show that $D(c)$ is a rational function which is not identically zero: let P be the set of the edges of a system of $p(U, X) = m$ disjoint paths between U and X. If we let $c_e = 1$ for $e \notin P$ and $c_e \to \infty$ for $e \in P$ then the defining equations cause the edges of P to shrink and the points of $f(U)$ belong to the gradually shrinking neighbourhood of the corresponding points of $f(X)$. Necessarily at some stage $\text{rank}(f(U)) = m$ since the points of $f(X)$ are affinely independent, and $D(C)$ is not identically zero.

Hence $D(c)$ vanishes only in a set of measure zero; we can have different such 'bad' set of coefficient-vectors for each $U \subset V$. But, since each of them is negligible, their union is negligible as well and we can find a coefficient-vector c which does not belong to any bad set.

□

Proof. (of Theorem 5.4.6) If G is k-connected then Lemma 5.4.8 implies the existence of a convex X-embedding f such that for every $Y \subset V$, $|Y| = k$, $f(Y)$ contains at least $p(X, Y) = k = |X|$ affinely independent points. Hence f is in general position. On the other hand, let f be the claimed embedding and X, Y arbitrary subsets of V of k elements. Then by Lemma 5.4.7, the number of affinely independent points of $f(Y)$ is at most $p(X, Y)$ but it is also at least $k = |X|$ since f is in general position.

□

5.5 Coin representations

Let G be a planar graph. A *coin representation* (CR) of G is a set of circles $\{C_v; v \in V(G)\}$ in the plane such that the interiors of the circles are pairwise disjoint and C_u, C_v touch if and only if G has the edge uv (see Figure 5.9). It has been proved by Koebe in 1936 that every planar graph has a CR. Clearly, each CR induces a convex embedding in the plane where each vertex is the center of the corresponding circle, and each edge is a straight line.

It will be convenient to consider also circle packings which contain one special circle, denoted by C_0, which behaves differently: it is required that none of the other circles intersects the exterior of C_0. We say that C_0 is *centered at infinity*. The corresponding 'planar embedding' has half-lines from the center of each circle C_v through $C_v \cap C_0$ (towards infinity). Let G be a connected planar graph. A *primal-dual CR* (PDCR) of G is a pair of simultaneous coin representations of G and the geometric dual G^* of the induced planar embedding of G such that for any pair of dual edges $e = uv, e^* = u^*v^*$, the circles C_u, C_v touch at the same point as the circles C_{u^*}, C_{v^*}. Moreover the two corresponding dual edges are perpendicular. We assume that the circle corresponding to the outer face of G is centered at infinity. The following theorem was proved by Brightwell and Scheinerman. For its proof as well as for the proof of its corollary see [MT].

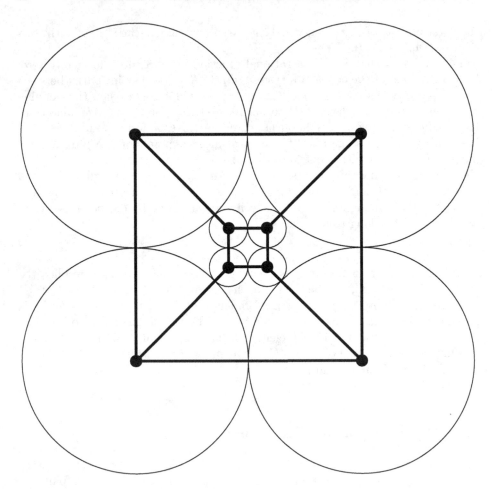

Figure 5.9. A coin representation of the cube

Theorem 5.5.1. *Let G be a 3-connected planar graph. Then G admits a PDCR representation.*

If Q is a convex polytope in $\mathbb{R}^3$ (i.e. a bounded set which is an intersection of finitely many half-spaces and not a subset of a line), then the *graph of polytope Q* has the corners of Q as its vertices, and the edges of Q as its edges.

Theorem 5.5.2. *If G is a 3-connected planar graph then there is a convex polytope Q in $\mathbb{R}^3$ whose graph is isomorphic to G, and such that all edges of Q are tangent to the unit sphere in $\mathbb{R}^3$.*

This leads to Steinitz's theorem:

Theorem 5.5.3. *A graph G is the graph of a convex polytope in $\mathbb{R}^3$ if and only if it is planar and 3-connected.*

Coin representations are closely related to *conformal (analytic) mappings* (see [MT]). Conformal mappings in turn are closely related to criticality of the Ising problem. Motivated by this connection, Mercat (see [MC]) defined *discrete analytic functions* on a simultaneous embedding of a finite graph and its dual in a 2-dimensional surface, and studied their criticality.

5.6 Counting fatgraphs: matrix integrals

Let M be an $N \times N$ matrix and let $f(M) = \sum_I a_I \prod_{(ij) \in I} M_{ij}$ be a polynomial in its entries, where I ranges over a finite system of multisets of elements of $N \times N$, and the a_I's are real constants. The basis of the *Wick game* is the following definition of $< f >$.

Definition 5.6.1. We let

$$< f > = \sum_I a_I < \prod_{(ij) \in I} M_{ij} > = \sum_I a_I \sum_P \prod_{(p,q) \in P} < M_p M_q >,$$

where P ranges over all partitions of I into pairs, and for $p = (p_1, p_2), q = (q_1, q_2)$ we have $< M_p M_q >$ non-zero only if $p_1 = q_2$ and $p_2 = q_1$ and in that case $< M_p M_q > = 1/N$.

A seminal technique of theoretical physics called Wick's theorem interprets the expression $< f >$ as a Gaußian matrix integral. The principally studied functions f are products of powers of the trace of M^k. For such functions f, $< f >$ has a useful graphic interpretation as the number of labeled fatgraphs (maps) with given degree sequence, sorted by their Euler characteristics (these maps are the *Feynman diagrams* for the matrix integral).

Let us define Gaußian integral and Gaußian matrix integral. We first consider the case $N = 1$. For an arbitrary real function f, the standard Gaußian integral is defined as

$$< f > = \frac{1}{\sqrt{2\pi}} \int_{-\infty}^{\infty} e^{-\frac{x^2}{2}} f(x) dx, \qquad (5.1)$$

where we abuse notation by a multiple use of the symbol $<>$. Note that $< 1 > = 1$. We are in particular interested in functions of the form $f(x) = x^{2n}$, where n is an integer. In order to compute $< x^{2n} >$, we introduce the so-called *source integral* $< e^{xs} >$ for a given real s. Taking the k-th derivative of $< e^{xs} >$ with respect to s and setting $s = 0$, we get

$$\frac{\partial^k}{\partial s^k} < e^{xs} > \Big|_{s=0} = \frac{1}{\sqrt{2\pi}} \int_{-\infty}^{\infty} e^{-\frac{x^2}{2}} \frac{\partial^k}{\partial s^k} e^{xs} \Big|_{s=0} dx$$

$$= \frac{1}{\sqrt{2\pi}} \int_{-\infty}^{\infty} e^{-\frac{x^2}{2}} x^k dx$$

$$= < x^k > . \qquad (5.2)$$

On the other hand the source integral becomes

$$
< e^{xs} > = \frac{1}{\sqrt{2\pi}} \int_{-\infty}^{\infty} e^{-\frac{x^2}{2}} e^{xs} dx
$$

$$
= \frac{1}{\sqrt{2\pi}} \int_{-\infty}^{\infty} e^{-\frac{(x-s)^2}{2} + \frac{s^2}{2}} dx
$$

$$
= e^{\frac{s^2}{2}} \frac{1}{\sqrt{2\pi}} \int_{-\infty}^{\infty} e^{-\frac{(x-s)^2}{2}} dx
$$

$$
= e^{\frac{s^2}{2}}. \tag{5.3}
$$

Thus we have

$$
< x^k > = \frac{\partial^k}{\partial s^k} < e^{xs} > \Big|_{s=0} = \frac{\partial^k}{\partial s^k} e^{\frac{s^2}{2}} \Big|_{s=0}. \tag{5.4}
$$

As a consequence, we obtain $< x^k > = 0$ for odd k, and $< x^2 > = 1$. Further since the derivative $\frac{\partial^{2n}}{\partial s^{2n}} e^{\frac{s^2}{2}} \Big|_{s=0}$ must be taken in pairs, $< x^{2n} >$ is the same as the number of ways to partition $2n$ elements into n pairs, which is $(2n-1)!! = (2n-1) \times (2n-3) \times \cdots 3 \times 1$.

Let us further consider the general case $N > 1$. We want to repeat the trick of the source integral, and as you will see below, for that we need to restrict ourselves to the set of matrices satisfying, for matrices M, S, that $\operatorname{tr} MS = \operatorname{tr} SM$. Hence we consider the set of the Hermitian matrices. Let $M = (M_{ij})$ be an $N \times N$ Hermitian matrix, i.e., $M_{ij} = \overline{M}_{ji}$ for every $1 \leq i, j \leq N$, where $\overline{M}_{ji}$ denotes the complex conjugate of M_{ji}. Let $dM = \prod_i dM_{ii} \prod_{i<j} dRe(M_{ij}) dIm(M_{ij})$ denote the standard Haar measure. This is the measure which enables integration over the Hermitian matrices. The Gaußian Hermitian matrix integral of an arbitrary function f is defined as

$$
< f(M) > = \frac{1}{Z_0(N)} \int e^{-N \operatorname{tr}(\frac{M^2}{2})} f(M) dM, \tag{5.5}
$$

where the integration is over all the $N \times N$ Hermitian matrices, and $Z_0(N)$ is the normalization factor making $< 1 > = 1$, i.e., $Z_0(N) = \int e^{-N \operatorname{tr}(\frac{M^2}{2})} dM$.

As before we are particularly interested in a function of the form $f(M) = \sum_I a_I \prod_{(ij) \in I} M_{ij}$, where I ranges over a finite system of multisets of elements of $N \times N$. We also introduce the source integral $< e^{tr(MS)} >$ for a given $N \times N$ Hermitian matrix S. It can easily be computed as

$$
< e^{tr(MS)} > = \frac{1}{Z_0(N)} \int e^{-N \operatorname{tr}(\frac{M^2}{2})} e^{tr(MS)} dM
$$

$$
= \frac{1}{Z_0(N)} \int e^{-N \operatorname{tr}(\frac{1}{2}(M - \frac{S}{N})^2)} e^{\frac{tr(S^2)}{2N}} dM
$$

$$
= e^{\frac{tr(S^2)}{2N}}, \tag{5.6}
$$

since the trace is linear and $\text{tr}(MS) = \text{tr}(SM)$, and thus we get

$$-N \ \text{tr}\left(\frac{M^2}{2}\right) + \text{tr}(MS) \ = \ -N \ \text{tr}\left(\frac{M^2}{2} - \frac{MS + SM}{2N}\right)$$

$$= \ -N \ \text{tr}\left(\frac{1}{2}\left(M - \frac{S}{N}\right)^2\right) + \frac{\text{tr}\left(S^2\right)}{2N}.$$

On the other hand, for any $1 \leq i, j \leq N$ we get

$$\frac{\partial}{\partial S_{ji}} e^{\text{tr}(MS)}\bigg|_{S=0} \ = \ \left(\frac{\partial}{\partial S_{ji}} \text{tr}(MS)\right) e^{\text{tr}(MS)}\bigg|_{S=0}$$

$$= \ \left(\frac{\partial}{\partial S_{ji}} \sum_{m,n} M_{mn} S_{nm}\right) e^{\text{tr}(MS)}\bigg|_{S=0}$$

$$= \ M_{ij}.$$

Thus the derivatives of the source integral becomes

$$\frac{\partial}{\partial S_{ji}} \frac{\partial}{\partial S_{lk}} \cdots < e^{\text{tr}(MS)} >\bigg|_{S=0}$$

$$= \ \frac{1}{Z_0(N)} \int e^{-N \ \text{tr}(\frac{M^2}{2})} \frac{\partial}{\partial S_{ji}} \frac{\partial}{\partial S_{lk}} \cdots e^{\text{tr}(MS)}\bigg|_{S=0} dM$$

$$= \ \frac{1}{Z_0(N)} \int e^{-N \ \text{tr}(\frac{M^2}{2})} M_{ij} M_{kl} \cdots dM$$

$$= \ < M_{ij} M_{kl} \cdots >. \tag{5.7}$$

Using (5.7) and (5.6), we obtain

$$< M_{ij} M_{kl} \cdots > \ \overset{(5.7)}{=} \ \frac{\partial}{\partial S_{ji}} \frac{\partial}{\partial S_{lk}} \cdots < e^{\text{tr}(MS)} >\bigg|_{S=0}$$

$$\overset{(5.6)}{=} \ \frac{\partial}{\partial S_{ji}} \frac{\partial}{\partial S_{lk}} \cdots e^{\frac{\text{tr}(S^2)}{2N}}\bigg|_{S=0} \tag{5.8}$$

and in particular

$$< M_{ij} M_{kl} > \ = \ \frac{\partial}{\partial S_{ji}} \frac{\partial}{\partial S_{lk}} e^{\frac{\text{tr}(S^2)}{2N}}\bigg|_{S=0}$$

$$= \ \frac{\partial}{\partial S_{ji}} \left(\frac{\partial}{\partial S_{lk}} \frac{\text{tr}(S^2)}{2N}\right) e^{\frac{\text{tr}(S^2)}{2N}}\bigg|_{S=0}$$

$$= \ \frac{\partial}{\partial S_{ji}} \left(\frac{\partial}{\partial S_{lk}} \frac{\sum_{m,n} S_{mn} S_{nm}}{2N}\right) e^{\frac{\text{tr}(S^2)}{2N}}\bigg|_{S=0}$$

$$= \ \frac{\partial}{\partial S_{ji}} \frac{S_{kl}}{N} e^{\frac{\text{tr}(S^2)}{2N}}\bigg|_{S=0}$$

$$= \ \frac{\delta_{il}\delta_{jk}}{N}. \tag{5.9}$$

Further, it is clear that the derivatives in (5.8) and (5.9) must be taken in pairs (e.g., S_{ji} and S_{lk} with $l = i$ and $k = j$) to get a non-zero contribution. This yields *Wick's theorem*:

Theorem 5.6.2. *Let M and I be as above. Then*

$$< \prod_{(ij) \in I} M_{ij} > = \sum_{pairings, P \subset I^2} \prod_{((ij),(kl)) \in P} < M_{ij} M_{kl} >$$

$$= \sum_{pairings, P \subset I^2} \prod_{((ij),(kl)) \in P} \frac{\delta_{il} \delta_{jk}}{N},$$

where P ranges over all partitions of I into pairs.

Due to the linearity of the Gaußian matrix integeral, together with (5.10), we get that for a function f defined by $f(M) = \sum_I a_I \prod_{(ij) \in I} M_{ij}$,

$$< f > = \sum_I a_I < \prod_{(ij) \in I} M_{ij} > = \sum_I a_I \sum_P \prod_{(p,q) \in P} < M_p M_q > .$$

For $p = (p_1, p_2), q = (q_1, q_2)$ we have that $< M_p M_q >$ is non-zero only if $p_1 = q_2$ and $p_2 = q_1$, and in that case $< M_p M_q > = 1/N$. This is how we arrive at the definition of $< \cdot >$ in Definition 5.6.1.

Graphic interpretation. Next we will count fatgraphs (see Section 5.1). We recall that a fatgraph is a graph together with a set consisting of one cyclic permutation, of the incident edges, for each vertex. A fatgraph is *pointed* if for each vertex one incident edge is specified.

Observation 5.6.3. *Let F be a pointed fatgraph. The following procedure orients each face of F: for each vertex v, orient the first (clockwise) shore of each incident fatedge out of v, and the other shore towards v.*

The graphic interpretation for the non-zero contributions to $< f >$ where $f(M) = \sum_I a_I \prod_{(ij) \in I} M_{ij}$, is as follows. We represent M_{ij} as a half-fatedge consisting of two end points and two lines with the opposite orientation such that i is associated with the out-going line and j with the in-coming line: Further,

Figure 5.10. A half-fatedge

(5.9) can be interpreted as the fact that two half-fatedges M_{ij} and M_{kl} construct a fatedge with oppositely oriented shores and with weight $1/N$ if and only if $i = l$ and $j = k$: A fatedge with oppositely oriented shores will be called a *decorated fatedge* (see Figure 5.11).

$$< M_{ij}, M_{kl} >= \frac{1}{N} \qquad \longleftrightarrow \qquad \begin{matrix} i \bullet \!\!\longrightarrow\!\! \bullet \; l, & l = i \\ j \bullet \!\!\longleftarrow\!\! \bullet \; k, & k = j \end{matrix}$$

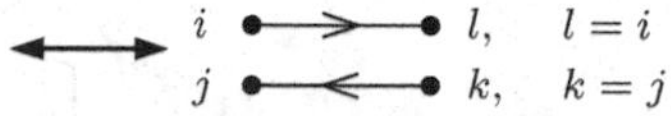

Figure 5.11. A decorated fatedge

For example, let us consider $f(M) = \mathrm{tr}(M^n)$. By the definition of the trace we get

$$\mathrm{tr}(M^n) = \sum_{1 \le i_1, i_2, \cdots, i_n \le N} M_{i_1 i_2} M_{i_2 i_3} \cdots M_{i_n i_1}.$$

Following the above graphic interpretation we represent $\mathrm{tr}(M^n)$ as a star fat diagram with n decorated half- fat edges arranged in a clockwise *pointed* order and such that for each half- fat edge, its first shore (clockwise along the center) is oriented from the center, as in Figure 5.12. Moreover, using the matrix Wick

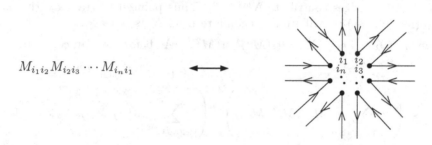

$$M_{i_1 i_2} M_{i_2 i_3} \cdots M_{i_n i_1}$$

Figure 5.12. $\mathrm{tr}(M^n)$ and its graphic interpretation as a star fat diagram

theorem we can compute

$$< \mathrm{tr}(M^n) > \; = \; < \sum_{1 \le i_1, i_2, \cdots, i_n \le N} M_{i_1 i_2} M_{i_2 i_3} \cdots M_{i_n i_1} >$$

$$= \sum_{1 \le i_1, i_2, \cdots, i_n \le N} \; \sum_{pairing} \prod < M_{i_k i_{k+1}} M_{i_l i_{l+1}} >$$

$$= \sum_{1 \le i_1, i_2, \cdots, i_n \le N} \; \sum_{pairing} \prod \frac{\delta_{i_k i_{l+1}} \delta_{i_l i_{k+1}}}{N}. \qquad (5.10)$$

Note that n should be even in order to get a non-zero contribution to (5.10) and thus we set $n = 2m$. Further, observe that only a fraction of possible pairings have a non-zero contribution to (5.10); we can think of a pairing as a pointed fatgraph with one island, whose faces are oriented as in Observation 5.6.3. It indeed defines uniquely an embedding on a surface (see Figure 5.13). Let F be a contributing pointed fatgraph. Certainly it has $n/2 = m$ edges. Since each edge contibutes $1/N$ to (5.10), each pairing gets $1/N^m$ from all its

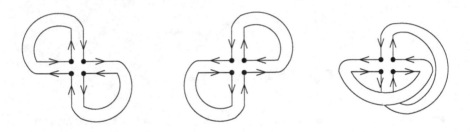

Figure 5.13. All possible fatgraphs with one island and $n = 4$

edges. However, we should count the contributions due to the summations over $1 \leq i_1, i_2, \cdots, i_n \leq N$. We notice that in each pairing with non-zero contribution, each (oriented) face attains independently exactly one index from 1 to N. The faces thus contribute $N^{p(F)}$ to (5.10); we recall that $p(F)$ denotes the number of faces of F. Summarizing, each pointed fatgraph F with one island and m edges contribute $N^{p(F)-m}$. Thus pointed fatgraphs with genus zero (i.e., planar) contribute the leading term in N as $N \to \infty$.

Example 5.6.4. Let $f(M) = \text{tr}(M^3)^4 \, \text{tr}(M^2)^3$. As before we can compute

$$< \text{tr}(M^3)^4 \, \text{tr}(M^2)^3 > =$$

$$< \left(\sum_{1 \leq i_1, i_2, i_3 \leq N} M_{i_1 i_2} M_{i_2 i_3} M_{i_3 i_1} \right)^4 \left(\sum_{1 \leq j_1, j_2 \leq N} M_{j_1 j_2} M_{j_2 j_1} \right)^3 > \quad (5.11)$$

Analogously to the previous example, (5.11) equals $\sum_F N^{p(F)-e(F)}$, where the sum is over all pointed fatgraphs F (not necessarily connected) cosisting of four islands of degree 3, and three islands of degree 2 (see Figure 5.14).

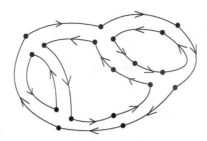

Figure 5.14. A fatgraph interpreted from $< \text{tr}(M^3)^4 \, \text{tr}(M^2)^3 >$

Summarizing, if $f(M)$ is a product of the traces of powers of M then $< f >$ equals the number of pointed fatgraphs F with the degree sequence given by the powers, and weighted by $N^{p(F)-e(F)}$.

This machinery is useful for instance for obtaining a formula for the number of fatgraphs with a given degree sequence, sorted by their genus. Next we explain this.

We first define, for a formal power series X, the exponential function $\exp(X)$, or e^X, by the formula

$$e^X = \sum_{n=0}^{\infty} \frac{X^n}{n!} = 1 + X + \frac{X^2}{2!} + \frac{X^3}{3!} + \frac{X^4}{4!} + \cdots .$$

As usual, the inverse (in the ring of the formal power series) of the exponential function is called the logarithm. For a formal power series X, the logarithm $\log(X)$ is defined by the formula

$$\begin{aligned} \log(X) &= \sum_{n=1}^{\infty} \frac{(-1)^{n+1}}{n}(X-1)^n \\ &= (X-1) - \frac{(X-1)^2}{2} + \frac{(X-1)^3}{3} - \frac{(X-1)^4}{4} \cdots . \end{aligned}$$

We consider a function ψ which maps each $N \times N$ Hermitian matrix M to

$$\psi(M) := e^{(-N)\sum_{i\geq 1} z_i \operatorname{tr}\left(\frac{M^i}{i}\right)}. \tag{5.12}$$

Finally, for a formal power series $X = \sum_{n\geq 0} a_n z^n$ with the coefficients in the ring of the $N \times N$ Hermitian matrices we define $< X >= \sum_{n\geq 0} < a_n > z^n$. Taking the logarithm of $< \psi >$ we get the following formula for connected maps sorted by their genus. For this and other interesting applications see the survey of DiFrancesco [F]. Let us remark here that in the applications of the formal power series, generating functions of connected objects are often expressed as the logarithm.

Theorem 5.6.5.

$$\log < \psi >= \sum_{g\geq 0} N^{2-2g} \sum_{(n_1,\cdots,n_k)\in\mathbb{N}^k} \prod_{i=1}^{k} \frac{(-z_i)^{n_i}}{n_i!} M_g(n_1,\cdots,n_k)$$

where $M_g(n_1,\cdots,n_k)$ denotes the number of connected fatgraphs with genus g and n_i vertices of degree i for $1 \leq i \leq k$.

The equality in Theorem 5.6.5 holds only on the level of *formal power series*. The convergence of the formal power series in Theorem 5.6.5 was studied extensively. For example, the following theorem appears in [EM].

Theorem 5.6.6. *Let a function $\tilde{\psi}$ map each $N \times N$ Hermitian matrix M to*

$$\tilde{\psi}(M) := e^{(-N)\sum_{i\in I} z_i \operatorname{tr}\left(\frac{M^i}{i}\right)},$$

where I is a finite set of positive integers such that its maximum element is even. Then for each $i \in I$ there is $\epsilon_i > 0$ so that for z_i a non-zero real variable with $z_i \in (0, \epsilon_i)$, Theorem 5.6.5 holds as equalities between analytic functions (in variables z_i) when ψ is replaced by $\tilde{\psi}$.

Chapter 6

Game of dualities

6.1 Edwards-Anderson Ising model

A physical system consisting of many particles can be described on two levels:
Microscopically it is determined by its *configuration*, i.e., by the positions and
momenta of all particles. Knowing a configuration of a system which obeys the
laws of classical mechanics and which is not influenced from outside allows one in
principle to determine its exact configuration at any future time. Of course, the
configuration of a realistic large system cannot even approximately be known.
On the other hand, a good description of the macroscopic properties of such a
system is provided by a relatively small number of *observable parameters* like
total energy, temperature, entropy, etc. These macroscopic properties are mod-
elled, in mathematics, as parameters associated with probability distributions
on the space of all configurations.

This point of view was developed towards the end of the 19th century in the
works of Boltzmann and Gibbs. If a physical system that is confined to a finite
volume is not influenced from the outside, it is driven by its internal fluctuations
towards an *equilibrium distribution*, which maximises the entropy. The branch
of mathematics which studies equilibrium distributions is called ergodic theory.
Let us try to explain the basic setting by an elementary example, as in [KG].

Let X be a finite set, the (abstract) configuration space. Since X is finite, a
distribution p is given as an assignment of probability $p(x)$ to each $x \in X$. Let
us further assume $p(x) > 0$ for each $x \in X$. In ergodic theory, the distributions
on the configuration space are called *states*. We do not follow this convention
here; we reserve the name 'state' for a configuration of the Ising model.

The *entropy* of the distribution p is defined as $H(p) = -\sum_{x \in X} p(x) \log p(x)$.
This entropy, also called the *information entropy*, measures the amount of uncer-
tainty that the observer is left with when s(he) knows that the system follows
distribution p. This can be explained as follows. Let $A \subset X$. We want to
associate a real number I_A with it that can be interpreted as the amount of information in the claim '$z \in A$'. If one requires that I_A is a continuous function

of the probability $p(A) = \sum_{x \in A} p(x)$ and that $I_{A \cap B} = I_A + I_B$ for sets A, B such that $p(A \cap B) = p(A)p(B)$ (i.e. for independent events), the only possible choice is $I_A = -\log p(A)$, where the logarithm can be taken to any base. It is customary to use the natural logarithm. The entropy $H(p) = \sum_{x \in X} p(x) I_x$ thus models the average amount of information of the elementary claims $z = x$, for $x \in X$. In other words, $H(p)$ is the expected amount of information that can be gained from further observations on the system, if its present distribution is known to be p.

Let us continue specifying our model: each configuration $x \in X$ is assigned an *energy* $z(x) \in \mathbb{R}$. In the distribution p, the system has the *mean energy*

$$\mathbb{E}_p(z) = \sum_{x \in X} p(x) z(x).$$

The energy specifies the *partition function*

$$Z(\beta) = \sum_{x \in X} e^{-\beta z(x)}.$$

For $\beta \in \mathbb{R}$, the *Gibbs measure* p_β is defined by

$$p_\beta(x) = \frac{1}{Z(\beta)} e^{-\beta z(x)}.$$

Gibbs measures are characterized by the following *variational principle*:

Theorem 6.1.1. *Each Gibbs measure p_β with $\beta \in \mathbb{R}$ satisfies*

$$H(p_\beta) + \mathbb{E}_{p_\beta}(z) = \log Z(\beta) = \sup_p [H(p) + \mathbb{E}_p(-\beta z)].$$

A distribution p for which this supremum is attained is called an *equilibrium distribution*. Thus Gibbs measures are equilibrium distributions. In fact, p_β is the only equilibrium distribution for a given β.

In a physical context, if we set the Boltzmann constant $k_B = 1$ for simplicity, then $T = 1/\beta$ denotes the *temperature* of the system.

Free energy of the distribution p is defined by $F(p) = \mathbb{E}_p(z) - T H(p)$. We can reformulate the variational principle in the following way:

$$F(p) \geq F(p_{\frac{1}{T}}) = -T \log Z(\frac{1}{T}),$$

with equality if and only if $p = p_{\frac{1}{T}}$.

If the system is infinite, it is possible to define equilibrium distributions, but for given $\beta \in \mathbb{R}$ the equilibrium distribution need not be unique. For example, a coexistence of two equilibrium distributions (states) at a given temperature in the Ising model (described below) on an infinite graph is interpreted in statistical physics as a *phase transition* of the first order.

In more realistic physical models configurations usually have some spatial structure. A classical example is the *Ising model*. It was designed by Lenz around

1920 to explain ferromagnetism and was named after his student Ising who contributed to its theory. The idea is that iron atoms are situated at vertices of a graph $G = (V, E)$ and behave like small magnets that can be oriented upwards (have *spin* +1) or downwards (have *spin* −1). Physically, two magnets that are close to one another need less energy to be oriented in the same way than in the opposite way. This leads to the following simplified model:

We assume that each edge uv of G has an assigned weight (or, a *coupling constant*) $w(uv)$. A configuration of the system is an assignment of the *spin* $\sigma_v \in \{+1, -1\}$ to each vertex v. This describes the two possible spin orientations the vertex can take. As is custumary in statistical physics, we will call the configurations of the Ising model its *states*.

The *energy function* (or *Hamiltonian*) of the model is defined as

$$z(\sigma) = -\sum_{uv \in E} w(uv)\sigma_u\sigma_v.$$

We can rewrite the Hamiltonian in the following way:

$$z(\sigma) = \sum_{\{u,v\} \in C} w(uv) - \sum_{\{u,v\} \in E \setminus C} w(uv) = 2w(C) - W,$$

where C is the set of edges connecting the pairs of vertices of different spins, and $W = \sum_{\{u,v\} \in E} w(uv)$ is the sum of all edge weights in the graph. Clearly, if we find the value of max-cut (see Section 2.3 for the definition of max-cut, min-cut), we have found the maximum energy of the model. Similarly, min-cut corresponds to the minimum energy (groundstate). The partition function is defined by

$$Z(G, \beta) = \sum_{\sigma} e^{-\beta z(\sigma)}.$$

In the next section we describe an efficient algorithm to determine the groundstate energy of the Ising model for any planar graph. In fact the whole partition function may be determined efficiently for planar graphs, and the principal ingredient is the concept of enumeration dualities of Section 6.4 and Section 6.3.

6.2 Max-Cut for planar graphs

Observation 2.10.6 reduces, for planar graphs, the Max-Cut problem to the *Maximum even subset problem*. Given a graph $G = (V, E)$ with rational weights on the edges, find the maximum value of $\sum_{e \in E'} w(e)$ over all even subsets $E' \subset E$. The following theorem thus means that the Max-Cut problem is efficiently solvable for planar graphs.

Theorem 6.2.1. *The maximum even subset problem is efficiently solvable for general graphs.*

The theorem is proved by a reduction to the *matching problem*. An efficient algorithm to find a *maximum matching* of a graph has been found by Edmonds (see Section 2.6). Edmonds also found an efficient algorithm to solve the *weighted perfect matching problem*: Given a graph $G = (V, E)$ with rational weights on the edges, the weighted perfect matching problem asks for the maximum value of $\sum_{e \in E'} w(e)$ over all perfect matchings E' of G.

This algorithm together with the following reduction of Fisher proves Theorem 6.2.1.

Theorem 6.2.2. *Given a graph $G = (V, E)$ with a weight function w on E, it is possible to construct a graph $G' = (V', E')$ and a weight function w' on E' so that there is a natural weight preserving bijection between the set of the even sets of G and the set of the perfect matchings of G'.*

Proof. The graph G' may be constructed from G by a local transformation at each vertex, described in Figure 6.1. We let $w'(e) = w(e)$ for all edges e of G, and $w'(e) = 0$ for each new edge. It is straightforward to verify the statement

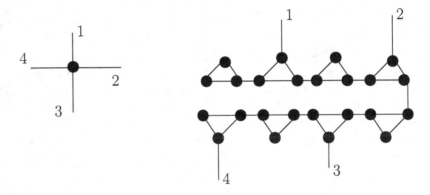

Figure 6.1. Each vertex is replaced by a path of triangles

of the theorem for G and G'.

□

In statistical physics, a perfect matching is known under the name of a *dimer arrangement* or a dimer configuration. The *dimer problem* is the problem to enumerate all perfect matchings, i.e., to find the *partition function*:

$$\mathcal{P}(G, x) = \sum_{E' \text{ perfect matching}} x^{\sum_{e \in E'} w(e)}.$$

In discrete mathematics, partition functions are called *generating functions*. The bijection of Theorem 6.2.2 does more than just a reduction of the maximum even set to the maximum perfect matching. The whole information about the even sets is transformed to the information about the perfect matchings. This may be expressed by the equality of the generating functions.

Let G be a graph and let w be a weight function on E. For $E' \subset E$ let $w(E') = \sum_{e \in E'} w(e)$. We define the generating functions $\mathcal{C}(G, x), \mathcal{E}(G, x), \mathcal{P}(G, x)$ of the edge-cuts, the even sets and the perfect matchings as follows.

$$\mathcal{C}(G, x) = \sum_{E' \text{ edge-cut}} x^{w(E')},$$

$$\mathcal{E}(G, x) = \sum_{E' \text{ even set}} x^{w(E')},$$

$$\mathcal{P}(G, x) = \sum_{E' \text{ perfect matching}} x^{w(E')}.$$

The transformation of Theorem 6.2.2 shows that $\mathcal{E}(G, x) = \mathcal{P}(G', x)$.

6.3 Van der Waerden's theorem

The Ising partition function for a graph G may be expressed in terms of the generating function of the even sets of the same graph G. This is a theorem of Van der Waerden. We will use the following standard notation: $\sinh(x) = (e^x - e^{-x})/2$, $\cosh(x) = (e^x + e^{-x})/2$, $\tanh(x) = \frac{\sinh(x)}{\cosh(x)}$.

Theorem 6.3.1.

$$Z(G, \beta) = 2^{|V|} \left(\prod_{uv \in E} \cosh(\beta w(uv)) \right) \left(\sum_{E' \subset E \text{ even}} \prod_{uv \in E'} \tanh(\beta w(uv)) \right).$$

Proof. We have

$$Z(G, \beta) = \sum_{\sigma} e^{\beta \sum_{uv} w(uv) \sigma_u \sigma_v} =$$

$$\sum_{\sigma} \prod_{uv \in E} (\cosh(\beta w(uv)) + \sigma_u \sigma_v \sinh(\beta w(uv))) =$$

$$\prod_{uv \in E} \cosh(\beta w(uv)) \sum_{\sigma} \prod_{uv \in E} (1 + \sigma_u \sigma_v \tanh(\beta w(uv))) =$$

$$\prod_{uv \in E} \cosh(\beta w(uv)) \sum_{\sigma} \sum_{E' \subset E} \prod_{uv \in E'} \sigma_u \sigma_v \tanh(\beta w(uv)) =$$

$$\prod_{uv \in E} \cosh(\beta w(uv)) \sum_{E' \subset E} (U(E') \prod_{uv \in E'} \tanh(\beta w(uv))),$$

where

$$U(E') = \sum_{\sigma} \prod_{uv \in E'} \sigma_u \sigma_v.$$

The proof is complete after noticing that $U(E') = 2^{|V|}$ if E' is even and $U(E') = 0$ otherwise. $\square$

It may help (optically) to shorten the right-hand-side of the formula by

$$\sum_{E' \subset E \text{ even}} \prod_{uv \in E'} \tanh(\beta w(uv))) = \mathcal{E}(G, x)|_{x^{w(uv)} := \tanh(\beta w(uv))},$$

where $\mathcal{E}(G, x)$ is the generating function of even subsets introduced at the end of Section 6.2. Let us recall that we may write the partition function as

$$Z(G, \beta) = \sum_{\sigma} e^{\beta \sum_{uv} w(uv) \sigma_u \sigma_v} = K \sum_{\sigma} e^{-2\beta \sum_{uv : \sigma(u) \neq \sigma(v)} w(uv)},$$

where $K = \sum_{\sigma} e^{\beta \sum_{uv} w(uv)}$ is a constant.

Hence the partition function $Z(G, \beta)$ may be looked at as the generating function of edge-cuts *with specified shores*. The theorem of Van der Waerden expresses it in terms of the generating function $\mathcal{E}(G, x)$ of the even sets of edges. We can also consider the honest *generating function of edge-cuts*

$$\mathcal{C}(G, x) = \sum_{E' \subset E \text{ edge-cut}} x^{w(E')}.$$

It turns out that $\mathcal{C}(G, x)$ may also be expressed in terms of $\mathcal{E}(G, x)$. This is a consequence of another theorem, of MacWilliams, which we explain now.

6.4   MacWilliams' theorem

Let $V = \mathbb{F}^n$ be a finite vector space over a finite field $\mathbb{F}$. Each subspace C of V of dimension k is called a *linear code of length n and dimension k*. If $\mathbb{F} = GF(2)$ then C is a *binary (linear) code*. The elements of a linear code are called *codewords*. The *weight* of a codeword is the number of its nonzero entries. The *weight distribution* of C is the sequence $A_0, A_1, \cdots, A_n$ where A_i equals the number of codewords of C of weight i, $0 \leq i \leq n$.

The *dual code* of C is denoted by C^* and consists of all the n-tuples $(d_1, \cdots, d_n)$ of $\mathbb{F}^n$ satisfying

$$c_1 d_1 + \cdots + c_n d_n = 0$$

in $\mathbb{F}$, for all codewords $(c_1, \cdots, c_n) \in C$. Hence, C^* is a code of length n and dimension $n - k$. The *weight enumerator* of C is the polynomial

$$A_C(t) = \sum_{i=0}^{n} A_i t^i.$$

MacWilliams' theorem for $\mathbb{F} = GF(2)$ reads as follows:

Theorem 6.4.1.

$$A_{C^*}(x) = \frac{1}{|C|}(1 + x)^n A_C \left(\frac{1 - x}{1 + x} \right).$$

The proof proceeds by a series of lemmas. We start by defining the *extended generating function of a code.*

Definition 6.4.2. Let $C \subset \{0,1\}^n$ be a binary linear code and $x = (x_1, \cdots, x_n)$, $y = (y_1, \cdots, y_n)$ be variables. The extended generating function of C is defined by $W_C(x,y) = \sum_{b \in C} W_b(x,y)$, where for $b = (b_1, \cdots, b_n)$, $W_b(x,y) = \prod_{i=1}^n W_{b_i}(x_i, y_i)$ and $W_{b_i}(x_i, y_i) = x_i$ if $b_i = 0$ and $W_{b_i}(x_i, y_i) = y_i$ if $b_i = 1$.

Lemma 6.4.3. *Let $b \in \{0,1\}^n$. Then*

$$W_b(x+y, x-y) = \sum_{c \in \{0,1\}^n} (-1)^{bc} W_c(x,y).$$

Proof. We note that

$$W_b(x+y, x-y) = \prod_{i=1}^n W_{b_i}(x_i + y_i, x_i - y_i) = \prod_{i=1}^n (x_i + (-1)^{b_i} y_i).$$

Expanding the right hand side we get a sum of 2^n terms of the form $\pm z_1 \cdots z_n$ where $z_i = x_i$ or $z_i = y_i$ and the sign is negative if and only if there is an odd number of indices i where $z_i = y_i$ and $b_i = 1$. Letting c index this sum of 2^n terms we get

$$\prod_{i=1}^n (x_i + (-1)^{b_i} y_i) = \sum_{c \in \{0,1\}^n} (-1)^{bc} \prod_{i=1}^n W_{c_i}(x_i, y_i).$$

$\square$

Lemma 6.4.4. *If $c \notin C^*$ then the sets $A_i = \{b \in C; cb = i\}$ have the same cardinality ($i = 0, 1$).*

Proof. We first note that both A_i are non-empty: $0 \in A_0$ and since $c \notin C^*$, there is $b \in C$ such that $bc = 1$. Let $b \in A_1$. Then $|b + A_0| = |A_0|$ and $b + A_0 \subset A_1$. Hence $|A_0| \leq |A_1|$. Analogously, $|-b + A_1| = |A_1|$ and $-b + A_1 \subset A_0$. Hence $|A_1| \leq |A_0|$.

$\square$

Lemma 6.4.5.
$$W_{C^*}(x,y) = \frac{1}{|C|} W_C(x+y, x-y).$$

Proof.
$$W_C(x+y, x-y) = \sum_{b \in C} \sum_{c \in \{0,1\}^n} (-1)^{bc} W_c(x,y) =$$

$$\sum_{b \in C} \sum_{c \in C^*} W_c(x,y) + \sum_{b \in C} \sum_{c \notin C^*} (-1)^{bc} W_c(x,y) = |C| W_{C^*}(x,y)$$

by Lemma 6.4.4.

$\square$

Lemma 6.4.6.
$$x^n A_C(y/x) = W_C(x, \ldots, x, y, \ldots, y).$$

Proof. We observe that

$$W_C(x, \ldots, x, y, \ldots, y) = x^n \sum_{b \in C} (y/x)^{w(b)},$$

where $w(b)$ denotes the weight of b.

$\square$

Proof. (of MacWilliam's Theorem 6.4.1) We have from Lemma 6.4.6 that

$$A_{C^*}(y) = W_{C^*}(1, \cdots, 1, y, \cdots, y).$$

Next we apply Lemma 6.4.5 and Lemma 6.4.6 again.

$\square$

We saw in section 2.3 that the set of the edge-cuts and the set of the even sets of edges form dual binary codes, hence MacWilliams' theorem applies here. Finally we remark that a version of MacWilliams' theorem is true more generally, for instance for linear codes over finite fields $GF(q)$.

6.5   Phase transition of 2D Ising

One of the basic observations about the world around us is that small changes of outside parameters may result in dramatic changes in some systems. An example is a phase transition from water to steam. Naturally, we want that our models of the reality also exhibit this feature. The 2-dimensional Ising model is one of the earliest models where a phase transition was proved to exist. But, how should the phase transitions (and the critical temperature) in the Ising model be defined? We saw in Section 6.1 that a phase transition of the first order exists if more than one equilibrium distribution (states) coexist. The *critical temperature* is defined as the supremum of the temperatures at which phase transitions of the first order occur.
We will assume in this section that a graph $G = (V, E)$ is a finite planar $n \times n$ square grid, and as is usual we denote by $N = n^2$ its number of vertices. Moreover for simplicity we will assume all the edges to have the same weight, i.e. we assume $w(uv) = J$ for each $uv \in E$. Let

$$Z(N, \beta) = Z(G, \beta) = \sum_{\sigma} e^{\beta J \sum_{uv \in E} \sigma_u \sigma_v}.$$

The *2-dimensional Ising model* lives in the infinite square grid. We may look at it as the limit of the Ising model on the square grid G for $N \to \infty$.
Let $F(\beta)$ be the *free energy*, i.e.

$$-F(\beta) = \lim_{N \to \infty} N^{-1} \log Z(N, \beta).$$

The critical temperature of the 2-dimensional Ising model is modeled as $T_c = 1/\beta_c$ such that at β_c, F is not a real analytic map of b (i.e., the Taylor series of F does not converge to F in β_c) .

In 1944, Onsager ([O]) provided a formula for the free energy of the 2-dimensional Ising problem. But three years earlier, Kramers and Wannier located its critical temperature, *under the assumption that the critical temperature exists and is unique.* Their beautiful argument ([KW]), based on a game of dualities, is retold now.

We will take advantage of the interplay between the geometric duality and the enumeration duality (Theorem 6.3.1). Let G^* denote the dual graph of G. A great property of the planar square grid is that it is essentially self-dual; there are some differences on the boundary, but who cares, we are playing anyway. We will **cheat** and assume that $G = G^*$. This omitting of boundary irregularities, for the planar square grid and for the hexagonal lattice in the next section, is the only cheating in the book, so enjoy it. If you wish to avoid the cheating, you can for example work with the toroidal grid instead of the planar square grid.

Low temperature expansion. Here we use the geometric duality. Let $\sigma : V \to \{1, -1\}$ be an Ising state in G. It corresponds to the assignments of $+$ or $-$ to the plaquettes (i.e. square faces) of G^*. An edge of G^* is called *frontal* for σ if it borders two plaquettes with opposite signs. It follows from the geometric duality that the set of frontal edges for σ is an even subset of G^*. Moreover, each even subset of edges of G^* corresponds to exactly two opposite Ising states in G. Summarizing,

$$Z(N, \beta) = \sum_{\sigma} e^{\beta J \sum_{uv \in E} \sigma_u \sigma_v} =$$

$$2 \sum_{E' \subset E^* \text{ even}} e^{-\beta J |E'| + \beta J |E^* - E'|} =$$

$$2e^{|E|\beta J} \sum_{E' \subset E \text{ even}} e^{-2|E'|\beta J},$$

where in the last equation we use $G^* = G$.

We recall that $\beta = 1/T$ where T denotes the temperature. If T goes to zero then β goes to infinity, and hence small even sets, i.e. short cycles, should dominate this expression of the partition function. This sort of explains why this formula is called the low temperature expansion.

High temperature expansion. Here we use Theorem 6.3.1. It gives

$$Z(N, \beta) = 2^N \cosh(\beta J)^{|E|} \sum_{E' \subset E \text{ even}} \tanh(\beta J)^{|E'|}.$$

If T goes to infinity then β goes to zero, and hence small cycles should again dominate this expression of the partition function.

Critical temperature of the 2-dimensional Ising model. Let us assume for simplicity that $J = 1$. At a critical point the free energy F is a non-analytic function of β. Moreover we *assume* that there is only one critical point. Then the expressions above help us locate it: let us define F^* by

$$F^*(x) = \lim_{N \to \infty} N^{-1} \log\Big(\sum_{E' \subset E \text{ even}} x^{|E'|} \Big).$$

Then, using the low and high temperature expansions, we get

$$-\beta F(\beta) = 2\beta + F^*(e^{-2\beta}) = \log(2 \cosh(\beta)) + F^*(\tanh(\beta)).$$

If we define β^* by $\tanh(\beta^*) = e^{-2\beta}$, we have

$$-\beta^* F(\beta^*) - \log(2 \cosh(\beta^*)) = F^*(\tanh(\beta^*)) = F'(e^{-2\beta}),$$

and hence

$$2\beta - \beta^* F(\beta^*) - \log(2 \cosh(\beta^*)) = F'(e^{-2\beta}) + 2\beta = -\beta F(\beta).$$

If β is large, β^* is small; the last equation relates the free energy at a low temperature to that at a high temperature. Hence, if there is only one critical value β_c, then necessarily $\beta_c = \beta_c^*$ which determines β_c: We have $\tanh(\beta_c) = e^{-2\beta_c}$, and hence $\sinh(2\beta_c) = 1$ and $\beta_c = 0.44068679....$

6.6   Critical temperature of the honeycomb lattice

Let us try to apply the same trick to the planar hexagonal (honeycomb) lattice H_{2N} with $2N$ vertices and edge-weights (coupling constants) L_1, L_2, L_3 defined by Figure 6.2. We have

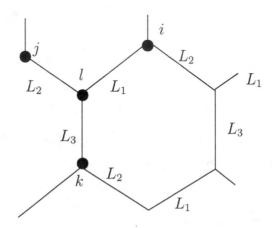

Figure 6.2. Weights in the hexagonal lattice

$$Z(H_{2N}, \beta) = \sum_{\sigma} e^{\beta(L_1 \sum_{il \in E_1} \sigma_i \sigma_l + L_2 \sum_{jl \in E_2} \sigma_j \sigma_l + L_3 \sum_{kl \in E_3} \sigma_k \sigma_l)},$$

where $E_i, i = 1, 2, 3$, denotes the set of the edges with the coupling constant L_i. If we disregard the boundary irregularities, the geometric dual of H_{2N} is the triangular lattice T_N with N vertices (see Figure 6.3). If we apply the high temperature expansion to H_{2N} and the low temperature expansion to T_N, we get an expression of $Z(H_{2N}, \beta)$ in terms of $Z(T_N, \beta)$:

Observation 6.6.1.

$$Z(T_N, \beta) = 2^N \prod_{i=1,2,3} \cosh(\beta L_i)^{|E_i|} \sum_{E' \subset E(T_N) \ even} \prod_{i=1,2,3} \tanh(\beta L_i)^{|E' \cap E_i|} =$$

$$2^{N-1} \prod_{i=1,2,3} \cosh(\beta L_i)^{|E_i|} Z(H_{2N}, \tanh(\beta)).$$

In order to extract the critical temperature, we need one more relation, and we will get it from the $\Delta - Y$ *(or, star-triangle) transformation.*

Definition 6.6.2. A $\Delta - Y$ (star-triangle) transformation in any graph consists in the exchange of a vertex l of degree 3 connected to independent vertices i, j, k (a Y), for three edges between vertices i, j, k that form a Δ (a triangle).

The star-triangle transformation is illustrated in the second part of Figure 6.3. We first note that the hexagonal lattice H_{2N} is bipartite; let the bipartition

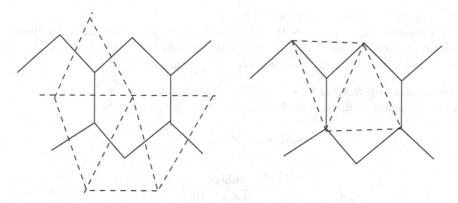

Figure 6.3. The honeycomb lattice and the associated triangular lattices formed by the geometric duality and by the star- triangle transformation

be $V = V_1 \cup V_2$. We will denote a typical vertex of V_2 by l (see Figure 6.2). The neighbours of l will be denoted by i, j, k as is in Figure 6.2. We also say that (i, j, k) is a triplet and we denote by $\mathcal{T}$ the set of all the triplets. The new trick is to apply the star-triangle transformation to all vertices of V_2.

The result is what we want, the triangular lattice T_N on the vertices of V_1. We can write

$$Z(H_{2N}, \beta) = \sum_{\sigma} e^{\beta(L_1 \sum_{il \in E_1} \sigma_i \sigma_l + L_2 \sum_{jl \in E_2} \sigma_j \sigma_l + L_3 \sum_{kl \in E_3} \sigma_k \sigma_l)} =$$

$$\sum_{\sigma: V_1 \to \{1, -1\}} \prod_{(i,j,k) \in \mathcal{T}} w(\sigma_i, \sigma_j, \sigma_k),$$

where

$$w(\sigma_i, \sigma_j, \sigma_k) = \sum_{\sigma_l \in \{1, -1\}} e^{\beta \sigma_l (L_1 \sigma_i + L_2 \sigma_j + L_3 \sigma_k)} = 2 \cosh(\beta(L_1 \sigma_i + L_2 \sigma_j + L_3 \sigma_k)).$$

Summarizing we can write $Z(H_{2N}, \beta)$ as

$$\sum_{\sigma: V_1 \to \{1, -1\}} \prod_{(i,j,k) \in \mathcal{T}} w(\sigma_i, \sigma_j, \sigma_k),$$

which may be regarded as a function on the triangular lattice T_N on V_1; its triangles are the triplets of $\mathcal{T}$.

We want to transform this further into the Ising partition function of this new triangular lattice T_N. Well, it is not a big deal: all we need is to compute constants R, K_1, K_2, K_3 so that for all values of $\sigma_i, \sigma_j, \sigma_k$ we have

$$w(\sigma_i, \sigma_j, \sigma_k) = R e^{K_1 \sigma_j \sigma_k + K_2 \sigma_k \sigma_i + K_3 \sigma_i \sigma_j};$$

for simplicity of presentation we let $\beta = 1$.

The constants K_1, K_2, K_3 are the edge-weights (coupling constants) for the triangular lattice, see Figure 6.4. It is straightforward to write this as a system of equations with variables R, K_1, K_2, K_3 and parameters L_1, L_2, L_3.

Observation 6.6.3. *Let $c = \cosh(L_1 + L_2 + L_3)$, $c_1 = \cosh(-L_1 + L_2 + L_3)$, and c_i is defined analogously for $i = 2, 3$. Then for $i = 1, 2, 3$,*

$$\sinh(2K_i) \sinh(2L_i) = k^{-1},$$

where

$$k = \frac{\sinh(2L_1) \sinh(2L_2) \sinh(2L_3)}{2(cc_1c_2c_3)^{1/2}}.$$

Moreover

$$R^2 = 2k \sinh(2L_1) \sinh(2L_2) \sinh(2L_3) = 2/(k^2 \sinh(2K_1) \sinh(2K_2) \sinh(2K_3)).$$

This expression for $Z(T_N, \beta)$ and Observation 6.6.1 suffice to extract the critical temperature for $Z(T_N, \beta)$.

The equations of Observation 6.6.3 lead to the famous *Yang-Baxter equation*. But before getting to it, let us explain the transfer matrix method first.

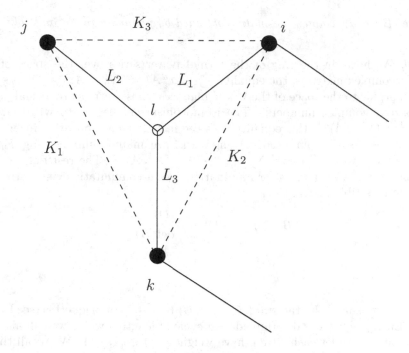

Figure 6.4. Dual weights

6.7 Transfer matrix method

Observation 6.7.1. *Let $D = (V, E)$ be a directed graph with weights $w(e)$ on its directed edges and let A be its $(V \times V)$ adjacency matrix, i.e. $A_{uv} = w(uv)$. Then the sum of the weights of the directed walks of D from u to v of length k is equal to $(A^k)_{uv}$.*

Proof. We proceed by induction on k. The case $k = 1$ is clear. For $k > 1$, any walk from u to v of length k consists of an edge uw and a walk from w to v of length $k - 1$. This is modeled by multiplication of matrices. $\square$

Hence the matrix A helps us to enumerate these walks. Let us further analyse the generating function

$$F_{uv}(x) = \sum_{n \geq 0} (A^n)_{uv} x^n,$$

for arbitrary $u, v \in V$.

Theorem 6.7.2.

$$F_{uv}(x) = \frac{(-1)^{u+v} \det(I - xA : v, u)}{\det(I - xA)},$$

where $(B : v, u)$ denotes the matrix obtained by removing row v and column u of B.

Proof. We have, in the ring of the formal power series over the space of the $V \times V$ complex matrices, the equality $\sum_{n \geq 0} x^n A^n = (Ix^0 - Ax)^{-1}$. This space is isomorphic to the space of the $V \times V$ matrices whose entries are formal power series over complex numbers. The isomorphism enables us to write $(Ix^0 - Ax)^{-1} = (I - xA)^{-1}$; this equality uses the fact that multiplicative inverses in any (not necessarily commutative) ring with 1 are unique. Summarizing, $F_{uv}(x)$ is the uv entry of the matrix $\sum_{n \geq 0} x^n A^n = (I - xA)^{-1}$. The rest follows from Cramer's rule, which holds for any matrix over a commutative ring with 1: If B is an invertible matrix then

$$(B^{-1})_{ij} = (-1)^{i+j} \frac{\det(B : j, i)}{\det(B)}.$$

$\square$

In the next example let the graph $G = (V, E)$ be a strip of plaquettes (see Figure 6.5). Each plaquette is determined by a cycle of length 4 which we will also call a plaquette. We let each edge e have weight $w(e) \in \{1, -1\}$. We recall that a *state* of the Ising model is any function σ from the set of the vertices to $\{1, -1\}$. We say that the edge $e = uv$ is *satisfied* by σ if $w(e)\sigma(u)\sigma(v) = 1$. A plaquette is *frustrated* if it has an odd number of edges e with weight $w(e) = -1$. No state can satisfy all the edges of a frustrated plaquette, and that is the reason for its name. We say that a state *satisfies* a plaquette if it satisfies the maximum possible number of its edges. Let us consider a graph G that is a ladder graph

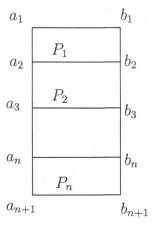

Figure 6.5. A strip of plaquettes

of frustrated plaquettes $P_1, ..., P_n$. We want to enumerate all states that satisfy all of these plaquettes; we will call them *satisfying states*. We start from the topmost pair of vertices a_1, b_1 and choose arbitrarily the spin from $\{1, -1\}$ for each of them. There are several ways in which we can choose the values for the next pair a_2, b_2, so that the plaquette $P_1 = (a_1, b_1, b_2, a_2)$ is satisfied. We have three options if the edge $\{a_1, b_1\}$ has been satisfied and only one option otherwise. It will be important how the spins of a_2, b_2 in a satisfying state depend on the spins of a_1, b_1. We encode this dependence in a 4×4 matrix A which is indexed by the spin assigments to $\alpha = (a_1, b_1)$ and $\beta = (a_2, b_2)$. There are four possibilities for each pair of spins and we will consider them as indices, in the order $(++, -+, +-, --)$. We let

- $a_{\beta\alpha} = 1$ if $[\alpha, \beta]$ satisfies the plaquette,

- $a_{\beta\alpha} = 0$ otherwise.

For example, consider a frustrated plaquette P where the edge $\{b_1, b_2\}$ has a negative weight while the other 3 edges have a positive weight. The corresponding matrix is

$$A = \begin{pmatrix} 1 & 0 & 1 & 1 \\ 0 & 0 & 0 & 1 \\ 1 & 0 & 0 & 0 \\ 1 & 1 & 0 & 1 \end{pmatrix}$$

The matrix A works as a linear operator. If we denote the unit vectors by e_α, then the vector Ae_α is the α-column of A. It describes the number of spin assignments, in the satisfying states, of vertices a_2, b_2, with the condition that the spins for a_1, b_1 are given by α. We can say that the β-entry of the vector Ae_α is equal to the number of satisfying states of the plaquette $P_1 = P$ where α gives the spins to a_1, b_1 and β gives the spins to a_2, b_2.

Let $A = A_1$ and let A_2 be the matrix describing the plaquette P_2. Let $\{a_3, b_3\}$ be the edge of P_2 disjoint from P_1. Then we can analogously say that the β-component of the vector $A_2 A_1 e_\alpha$ is equal to the number of satisfying states of the column of two plaquettes P_1, P_2 where α gives the spins to a_1, b_1 and β gives the spins to a_3, b_3. Inductively, for a strip of n plaquettes we get that

$$g(\alpha, \beta) = e_\beta A_n A_{n-1} \cdots A_1 e_\alpha$$

is the number of satisfying states of the strip of plaquettes $P_1, ..., P_n$, with the spins on the topmost pair a_1, b_1 given by α and the spins on the bottom pair a_{n+1}, b_{n+1} given by β.

In case the strip of plaquettes is periodic, we just identify the top with the bottom and we get that the total number of satisfying states is

$$g = \sum_\alpha e_\alpha^T A_n A_{n-1} \cdots A_1 e_\alpha = \mathrm{tr}(A_n...A_1).$$

For example, let us consider a periodic strip of n equal frustrated plaquettes characterized by the matrix A above. Then the number of satisfying states is

$$g = \mathrm{tr}(A^n) = \sum_{i=1}^{4} \lambda_i^n,$$

where $\lambda_1, \lambda_2, \lambda_3, \lambda_4$ are the eigenvalues of A. Since the largest eigenvalue of A is $\lambda_1 = 1 + \sqrt{2}$ (and the other eigenvalues are ± 1 and $1 - \sqrt{2}$), the number of satisfying states grows approximately as $(1 + \sqrt{2})^n$. Note that the matrix A in our example is symmetric, which not need be true in general. The singular values rather than the eigenvalues are then useful in the analysis.

It will help us in the next section to summarize what we are doing here. We want to calculate the number of satisfying states of a (periodic) strip of plaquettes. We consider a row of two spins σ_1, σ_2, and for each plaquette we consider an operator which is a $2^2 \times 2^2$ matrix that encodes the satisfying assignments of the plaquette. The product of these operators applied to the vector (σ_1, σ_2) then describes the total number of satisfying states of the whole strip, when the topmost spins are given by σ_1, σ_2.

It is intuitive to think that *we are building* the strip of plaquettes, rather than building the operator that describes the number of satisfying states.

6.8   The Yang-Baxter equation

As illustrated by the previous example, in two-dimensional lattice problems it is often useful to consider a row of spins $\sigma_1, \cdots, \sigma_m$ ($m = 2$ in the strip of plaquettes), and operators that are built up gradually by adding contributions of small parts (plaquettes in the previous example). These operators are $2^m \times 2^m$ matrices, with rows labelled by $(\sigma_1, \cdots, \sigma_m)$ and columns labelled by $(\sigma_1', \cdots, \sigma_m')$. Let us now build in this way an operator that describes the Ising partition function of the $m \times m$ square grid. The smallest building blocks are the following basic operators $s_1, \cdots, s_m$ and $c_1, \cdots, c_m$. Operator s_i is the diagonal matrix with entries σ_i and c_i is the operator that reverses the spin in position i, i.e.

$$(c_i)_{\sigma\sigma'} = \delta(\sigma_1, \sigma_1') \cdots \delta(\sigma_{i-1}, \sigma_{i-1}')\delta(\sigma_i, -\sigma_i')\delta(\sigma_{i+1}, \sigma_{i+1}') \cdots \delta(\sigma_m, \sigma_m'),$$

Where δ is the Kronecker's delta function defined in the introduction. Writing $[x, y] = xy - yx$ for the commutator of the operators x, y, we have $s_i^2 = c_i^2 = I$, $s_i c_i + c_i s_i = 0$, and $[s_i, s_j] = [c_i, s_j] = [c_i, c_j] = 0$ for $i \neq j$.

In the Ising model, the building operators are

$$P_1(K), \cdots, P_{m-1}(K), Q_1(L), \cdots, Q_m(L),$$

where

$$[P_i(K)]_{\sigma\sigma'} = e^{K\sigma_i\sigma_{i+1}}\delta(\sigma_1, \sigma_1') \cdots \delta(\sigma_m, \sigma_m'),$$

$$[Q_i(K)]_{\sigma\sigma'} = e^{L\sigma_i\sigma_i'}\delta(\sigma_1, \sigma_1') \cdots \delta(\sigma_{i-1}, \sigma_{i-1}')\delta(\sigma_i, \sigma_i')\delta(\sigma_{i+1}, \sigma_{i+1}') \cdots \delta(\sigma_m, \sigma_m').$$

The effect of the operator $P_i(K)$ is to add the contribution of an edge with coupling constant K, between vertices i and $i+1$. The effect of $Q_i(L)$ is to introduce a new vertex in position i, and add the contribution of the edge between the new and old vertices in position i, with interaction coefficient L. The operator $P_i(K)$ thus adds a horizontal edge and the operator $Q_i(L)$ adds a vertical edge. We can also describe by means of operators the contribution of triangles and stars to the Ising partition function. For that, let us define operators $U_1, \cdots, U_{2m-1}$ by

$$U_{2j}(K, L) = P_j(K),$$

$$U_{2j-1}(K, L) = (2\sinh(2L))^{-1/2}Q_j(L).$$

The equations of Observation 6.6.3 imply

$$U_{i+1}(K_1, L_1)U_i(K_2, L_2)U_{i+1}(K_3, L_3) = U_i(K_3, L_3)U_{i+1}(K_2, L_2)U_i(K_1, L_1),$$

for $i = 1, \cdots, 2N - 2$. If i is even then the LHS of the equation above adds the contribution of the corresponding star with the coupling constants (L_1, L_2, L_3) and the RHS adds the contribution of the corresponding triangle with the coupling constants (K_1, K_2, K_3). This is reversed if i is odd.
The equation

$$U_{i+1}U_iU_{i+1} = U_iU_{i+1}U_i$$

is known as the *Yang-Baxter equation* and we get back to it when we will speak about knots.

Chapter 7

The zeta function and graph polynomials

7.1   The Zeta function of a graph

The theory of the Möbius function connects the principle of inclusion and exclusion (PIE) with another basic concept, namely the *zeta function of a graph*. In this section we discuss the theorem of Bass which will be useful in Chapter 9, and the MacMahon Master theorem, which will be useful in Chapter 8. Then we study graph polynomials, which constitute the basic connection between enumeration and the partition functions of statistical physics.

Let $G = (V, E)$ be a graph. If $e \in E$ then we let a_e denote an orientation of e (arbitrary but fixed), and a_e^{-1} the reversed directed edge to a_e. A circular sequence $p = v_1, a_1, v_2, a_2, ..., a_n, v_{n+1}$ and $v_{n+1} = v_1$ is called a *prime reduced cycle* if the following conditions are satisfied: $a_i \in \{a_e, a_e^{-1} : e \in E\}$, $a_i \neq a_{i+1}^{-1}$ and $(a_1, ..., a_n) \neq Z^m$ for some sequence Z and $m > 1$.

Definition 7.1.1. Let $G = (V, E)$ be a graph. The Ihara-Selberg function of G is

$$I(u) = \prod_{\gamma}(1 - u^{|\gamma|})$$

where the product is defined by

$$\prod_{\gamma}(1 - u^{|\gamma|}) = \sum_{\mathcal{G}}(-1)^{|\mathcal{G}|}u^{\sum_{\gamma \in \mathcal{G}}|\gamma|},$$

and the sum is over all finite sets $\mathcal{G}$ of the prime reduced cycles. The zeta function of G is

$$Z(u) = I(u)^{-1}.$$

The theorem of Bass reads as follows:

Theorem 7.1.2. *(Bass' theorem) For any graph G*

$$I(u) = \det(I - uT),$$

where T is the matrix of transitions between directed edges defined as follows: Let $a, a' \in \{a_e, a_e^{-1} : e \in E\}$. If the terminal vertex of a is the initial vertex of a' and $a' \neq a^{-1}$ then $T_{a,a'} = 1$, otherwise $T_{a,a'} = 0$.

Next we write down the MacMahon Master theorem.

Theorem 7.1.3. *(MacMahon Master theorem) Let $A = (a_{ij})$ be an $n \times n$ matrix, and let $x = (x_1, \cdots, x_n)$ be a vector of commuting variables. The coefficient of $x_1^{m_1} \cdots x_n^{m_n}$ in*

$$\prod_{i=1}^{n}(\sum_{j=1}^{n} a_{ij}x_j)^{m_i}$$

is equal to the coefficient of $x_1^{m_1} \cdots x_n^{m_n}$ in the expansion of $[\det(I - xA)]^{-1}$.

We include the proofs of these theorems based on the theory of Lyndon words. We closely follow [FZ]. Let X be a non-empty linearly ordered set, and consider the set X^* of all finite words from X. Let $<$ denote the lexicographic ordering on X^* derived from the linear ordering on X: for $u \neq v$ we say that $u < v$ if $v = uz$ for some $z \in X^*$, or $u = ras, v = rbt$ with $a < b$ and $r, s, t \in X^*$. We consider the set X^* of all words from X equipped with the binary operation of concatenation:

$$(a_1, \ldots, a_n)(b_1, \ldots, b_m) = (a_1, \ldots, a_n, b_1, \ldots, b_m).$$

A *Lyndon word* is a nonempty word of X^* that is prime (i.e., it cannot be written as a power of a shorter word), and minimal among its cyclic rearrangements (for example, 221 is not a Lyndon word since $221 > 122$. Let $\mathcal{L}$ denote the set of all Lyndon words.

Observation 7.1.4. *A non-empty word w is Lyndon if and only if w is smaller than any of its proper right factors if and only if $w \in X$ or $w = lm$ with $l, m \in \mathcal{L}$ and $l < m$.*

The following theorem is called Lyndon's factorization theorem.

Theorem 7.1.5. *Each nonempty word $l \in X^*$ can be uniquely written as a nonincreasing concatenation of Lyndon words: $l = l_1 l_2 \cdots l_n$, $l_k \in \mathcal{L}$, $l_1 \geq l_2 \geq \cdots \geq l_n$.*

Proof. To prove the theorem, we simply take a factorization $l = l_1 l_2 \cdots l_n$ into Lyndon words (a factorization like that clearly exists since each element of X is a Lyndon word) such that n is as small as possible. The Lyndon words in this factorization must be nonincreasing by Observation 7.1.4. The observation also proves the uniqueness.

$\square$

Next we consider formal power series with integer coefficients, and with variables in X, which are not commuting. It is convenient to use the symbol X^* to denote $\sum_{l \in X^*} l$. As an exercise in this notation (we denote by X_r^* the set of the reversed words of X^*) prove that the Lyndon factorization theorem is the same as

$$\prod_{l \in \mathcal{L}} (1 - l)^{-1} = X_r^* = X^* = (1 - \sum_{z \in X} z)^{-1},$$

where the indices in the product appear in the increasing order.

We get Amitsur's identity as a useful corollary:

Proposition 7.1.6. *Let X be the set of matrices $A_1, \ldots, A_k$, linearly ordered by their indices. Then*

$$\det(I - (A_1 + \cdots + A_k)) = \prod_{l \in \mathcal{L}} \det(I - l).$$

Proof. We can write as above

$$\prod_{l \in \mathcal{L}} (I - l)^{-1} = (I - (\sum_{z \in X} z))^{-1}.$$

Now we take the inverse of this identity, and take the determinant of both sides. This finishes the proof. $\square$

Let $\mathcal{B}$ be an $X \times X$ matrix whose entries are commuting variables. We denote the ij-entry of $\mathcal{B}$ by $b(i, j)$. We can think of $b(i, j)$ as the weight of the *transition* between the elements i, j of X.

Definition 7.1.7. Let $w = x_1 x_2 \cdots x_m$ be a nonempty word of X^*. We define

$$\beta_{circ}(w) = b(x_1, x_2) b(x_2, x_3) \cdots b(x_{m-1}, x_m) b(x_m, x_1),$$

and $\beta_{circ}(w) = 1$ if w is empty. Let $w = l_1 l_2 \cdots l_n$ be the expression of w as the nonincreasing concatenation of Lyndon words. We further define

$$\beta_{dec}(w) = \beta_{circ}(l_1) \beta_{circ}(l_2) \cdots \beta_{circ}(l_n).$$

Finally, when the m letters of w are written in the *nondecreasing order*, we get the word $w' = x_1' x_2' \cdots x_m'$. We let

$$\beta_{vert}(w) = b(x_1', x_1) b(x_2', x_2) \cdots b(x_m', x_m).$$

We also let $\beta_{dec}(w) = \beta_{vert}(l) = 1$ if w is empty.

The following elementary observation is an exercise in the use of these new notions; it will be useful.

Observation 7.1.8. *Let $w \in X^*$ and let $w = l_1 \cdots l_n$ be the decomposition into a nonincreasing sequence of Lyndon words. Further, let $w = d_1 \cdots d_r$ be the decreasing factorization, where each new factor starts always when a letter smaller than or equal to each letter to its left appears. Then each Lyndon word l_i is a concatenation of factors d_j. Moreover*

$$\beta_{dec}(w) = \beta_{circ}(l_1)\beta_{circ}(l_2) \cdots \beta_{circ}(l_n) = \beta_{circ}(d_1)\beta_{circ}(d_2) \cdots \beta_{circ}(d_r).$$

The following theorem summarizes the relations among the notions we introduced. Both the theorem of Bass and the MacMahon Master theorem are straightforward consequences.

Theorem 7.1.9. *The following properties hold.*

$$\prod_{l \in \mathcal{L}}(1 - \beta_{circ}(l))^{-1} = \sum_{w \in X^*} \beta_{dec}(w) \tag{1}$$

$$\sum_{w \in X^*} \beta_{dec}(w) = \sum_{w \in X^*} \beta_{vert}(w) \tag{2}$$

$$\sum_{w \in X^*} \beta_{vert}(w) = (\det(I - \mathcal{B}))^{-1} \tag{3}$$

$$\prod_{l \in \mathcal{L}}(1 - \beta_{circ}(l)) = \det(I - \mathcal{B}) \tag{4}$$

Proof of the MacMahon Master theorem and Bass's theorem. The MacMahon Master theorem follows from statement (3) of Theorem 7.1.9. Bass's theorem is the statement (4) of Theorem 7.1.9 for X equal to the orientations of the edges, and $b(e, e') = u$ if e is a successor of e' and e is not the reversed e'.

Proof. (of Theorem 7.1.9) First note that (1),(2) and (4) imply (3). Next let us associate, with each Lyndon word l, a variable denoted by $[l]$. We assume that these variables are distinct and commute with each other. Let $\beta([l]) = \beta_{circ}(l)$. We have

$$\prod_{l \in \mathcal{L}}(1 - \beta_{circ}(l))^{-1} = \prod_{l \in \mathcal{L}}(1 - \beta([l]))^{-1} = \sum_{[l_{i_1}], \cdots, [l_{i_n}]} \beta([l_{i_1}])\beta([l_{i_2}]) \cdots \beta([l_{i_n}]),$$

where the sum is over all the commuting monomials $[l_{i_1}], \cdots, [l_{i_n}]$, or equivalently over the nonincreasing collections $l_{i_1} \geq \cdots \geq l_{i_n}$ of Lyndon words. By Theorem 7.1.5, this equals

$$\sum_{w \in X^*} \beta_{dec}(w).$$

This proves (1).

In order to prove (2), we construct a bijection f of X^* onto itself so that for each w, $f(w)$ is a rearrangement of w and $\beta_{dec}(w) = \beta_{vert}(f(w))$. The construction goes as follows: Let $w \in X^*$ and let $w = l_1 \cdots l_n$ be the decomposition into the nonincreasing sequence of Lyndon words, and let $w = d_1 \cdots d_r$ be the decreasing factorization of w (see Observation 7.1.8). We define a set S of ordered pairs as follows: for each $1 \leq i \leq r$, if $d_i = i_1 \cdots i_p$ then we put the pairs $(i_1, i_2), \cdots, (i_{p-1}, i_p), (i_p, i_1)$ into S. We define $f(w)$ to be the word consisting of the second elements of each pair of S, written according to the nondecreasing lexicographic order of S. The properties of f follow from Observation 7.1.8.

Finally we show that (4) follows from Amitsur's identity (Theorem 7.1.6). We consider the lexicographic order on the indices of $\mathcal{B}$ (i.e. on the elements of $X \times X$). If ij is the m-th pair then let A_m be the matrix whose entries are all zero except $(A_m)_{ij} = b(i, j)$. Then $A_1 + \cdots + A_{|X|^2} = \mathcal{B}$.

Consider a word $l = (i_1, j_1), \cdots, (i_p, j_p)$ in the alphabet X^2 and let $A_l = \prod_{s=1}^{p} A_{(i_s, j_s)}$. If $j_1 = i_2, j_2 = i_3, \cdots, j_{p-1} = i_p$ then A_l is the matrix whose elements are all zero except $(A_l)_{i_1 j_p} = b(i_1, i_2) b(i_2, i_3) \cdots b(i_p, j_p)$. In all other cases A_l is the zero matrix. Hence, if $j_p = i_1$ we have $\det(I - A_l) = 1 - b(i_1, i_2) b(i_2, i_3) \cdots b(i_p, i_1)$, and in all the other cases we have $\det(I - A_l) = 1$. It means that the infinite product in Amitsur's identity may be restricted to the Lyndon words $l = (i_1, j_1), \cdots, (i_p, j_p)$ satisfying $j_1 = i_2, j_2 = i_3, \cdots, j_{p-1} = i_p, j_p = i_1$. But these are in bijection with the Lyndon words $i_1 \cdots i_p$ in the alphabet X.

$\square$

We conclude this section by a reformulation of the MacMahon Master theorem in terms of flows. A *natural flow* f on a digraph G is a function $f : E \longrightarrow \mathbb{N}$ on the edges of G that satisfies Kirchhoff's current law

$$\sum_{e \text{ begins at } v} f(e) = \sum_{e \text{ ends at } v} f(e)$$

at all vertices v of G. Let us set

$$f(v) = \sum_{e \text{ begins at } v} f(e).$$

Let $\mathcal{F}(G)$ denote the *set of all natural flows* of a digraph G. If β is a weight function on the set of edges of G and f is a flow on G, then

- the *weight* $\beta(f)$ of f is given by $\beta(f) = \prod_e \beta(e)^{f(e)}$, where $\beta(e)$ is the weight of the edge e.

- The *multiplicity* at a vertex v with outgoing edges $e_1, e_2, \cdots$ is given by $\text{mult}_v(f) = \binom{f(e_1) + f(e_2) + \cdots}{f(e_1), f(e_2), \cdots}$, and the *multiplicity* of f is given by $\text{mult}(f) = \prod_v \text{mult}_v(f)$.

- If E' is a subset of edges then we let $f(E') = \sum_{e \in E'} f(e)$.

Theorem 7.1.10. *If G is a digraph with the edge-weights given by matrix $\mathcal{B}$, then*

$$\frac{1}{\det(I - \mathcal{B})} = \sum_{f \in \mathcal{F}(G)} \beta(f) \, \text{mult}(f).$$

Proof. This is another reformulation of statement (3) of Theorem 7.1.9: we observe that $\beta_{vert}(w) = \beta(f)$ for a natural flow f and mult(f) elements $w \in X^*$.

$\qquad\qquad\qquad\qquad\qquad\qquad\qquad\qquad\qquad\qquad\qquad\qquad\qquad\qquad$ $\square$

For $r = 1$, the above corollary states that

$$\frac{1}{1-x} = \sum_{n=0}^{\infty} x^n$$

where $x = b_{11}$. Thus, Corollary 7.1.10 is a version of the geometric series summation.

7.2 Chromatic, Tutte and flow polynomials

In this section we introduce basic graph polynomials. In his paper [WH1] Whitney wrote down a formula for the number of ways of coloring a graph as one of the applications of the principle of inclusion and exclusion (PIE): Suppose we have a fixed number z of colors at our disposal. Any way of assigning one of these colors to each vertex of the graph in such a way that any two vertices which are joined by an edge are of different colors, will be called an admissible coloring, using z or fewer colors. We wish to find the number $M(G, z)$ of such admissible colorings. If there are n vertices in G, then there are z^n possible colorings. Let A_{uv} denote those colorings with the property that u and v are of the same color. Then the number of admissible colorings is, by PIE,

$$M(G, z) = \sum_{E' \subset E} (-1)^{|E'|} \left| \bigcap_{uv \in E'} A_{uv} \right|.$$

A typical term $|\bigcap_{uv \in E'} A_{uv}|$ is the number of ways of coloring (V, E') with z or fewer colors in such a way that any two vertices that are joined by an edge must be of the same color; thus all vertices in a single connected component are of the same color. If there are p connected pieces in (V, E'), the value of this term is therefore z^p. If there are (p, s) (this is Birkhoff's symbol) subgraphs of s edges in p components, the corresponding terms contribute the amount of $(-1)^s (p, s) z^p$ to $M(G, z)$. Therefore,

$$M(G, z) = \sum_{p,s} (-1)^s (p, s) z^p.$$

This function is called the *chromatic polynomial*. A basic recurrence relation for $M(G, z)$ is as follows: if $u, v \in V$ are such that $uv \notin E$ then in any proper coloring, u, v may receive the same color, or different colors. Hence,

$$M(G, z) = M(G/uv, z) + M(G + uv, z),$$

where G/uv denotes the graph obtained from G by identifying the vertices u and v, and $G + uv$ denotes the graph obtained from G by adding the edge uv. In other words, if $e \in E$ then

$$M(G, z) = M(G - e, z) - M(G/e, z).$$

For $A \subset E$ let $r(A) = |V| - c(A)$, where $c(A)$ denotes the number of components of (V, A). Then we can write

$$M(G, z) = z^{c(E)}(-1)^{r(E)} \sum_{A \subset E} (-z)^{r(E)-r(A)}(-1)^{|A|-r(A)}.$$

This leads directly to the Whitney rank generating function $S(G, x, y)$ defined by

$$S(G, x, y) = \sum_{A \subset E} x^{r(E)-r(A)} y^{|A|-r(A)},$$

and to the *Tutte polynomial*

$$T(G, x, y) = \sum_{A \subset E} (x-1)^{r(E)-r(A)}(y-1)^{|A|-r(A)}.$$

The Tutte polynomial enumerates many basic objects. As a first example, notice that for any connected graph G, $T(G, 1, 1)$ counts the number of spanning trees of G: indeed, the only non-zero terms of $T(G, 1, 1)$ are those for which $r(A) = r(E) = |A|$. These are exactly the spanning trees of G.

Theorem 7.2.1. *Let $G = (V, E)$ be a graph and $e \in E$. Then*
$S(G, x, y) = (x+1)S(G-e, x, y)$ *if e is a bridge,*
$S(G, x, y) = (y+1)S(G-e, x, y)$ *if e is a loop,*
$S(G, x, y) = S(G-e, x, y) + S(G/e, x, y)$ *otherwise.*
Furthermore, $S(G, x, y) = 1$ for any graph with no edges.

Proof. These formulas follow in a straightforward way by analyzing the definition of S. $\square$

Another important observation is the *universality* of Whitney and Tutte polynomials. It is expressed by the following theorem.

Theorem 7.2.2. *There is a unique map $\mathcal{M}$ from the class of finite graphs to the class of the integer polynomials in variables $x, y, \alpha, \beta, \gamma$ such that $\mathcal{M}(G) = \alpha^n$ if G is the graph of n vertices and no edges, and for every $e \in E(G)$ we have*
$\mathcal{M}(G) = x\mathcal{M}(G-e)$ *if e is a bridge,*
$\mathcal{M}(G) = y\mathcal{M}(G-e)$ *if e is a loop, and*
$\mathcal{M}(G) = \beta\mathcal{M}(G-e) + \gamma\mathcal{M}(G/e)$ *otherwise. Moreover,*
$\mathcal{M}(G) = \alpha^{c(G)}\beta^{|E|-r(G)}\gamma^{r(G)}T(G, \frac{\alpha x}{\gamma}, \frac{y}{\beta}).$

Proof. The uniqueness is immediate since the formulas provide a recursion for $\mathcal{M}$. It remains to observe that function $\alpha^{c(G)}\beta^{n(G)}\gamma^{r(G)}T(G, \frac{\alpha x}{\gamma}, \frac{y}{\beta})$ has the initial three properties. This follows directly from Theorem 7.2.1. $\square$

Next we define and study nowhere-zero flows.

Definition 7.2.3. Let G be a graph with a fixed orientation and let Z_k be the additive group of integers modulo k. A circulation on G with values in Z_k is called a Z_k-flow. Moreover if the circulation is never zero, then it is called a *nowhere-zero* Z_k-flow.

For an important illustration we consider an orientation of a $2-$edge-connected topological planar graph G which is *face-k−colorable*, i.e. its faces may be colored by k colors so that each edge belongs to the boundary of two faces of different colors. Consider the colors as elements of an additive group of order k, say Z_k. For each edge e let $r(e)$ and $l(e)$ be the colors of the faces on its left and right (G is directed!). Then $r − l$ is *nowhere-zero* and it is a *flow*: for each vertex v,

$$\sum_{(u,v)\in E} (r-l)((u,v)) = \sum_{(v,u)\in E} (r-l)((v,u)).$$

In fact, the following is true.

Theorem 7.2.4. *An orientation of a $2−$edge-connected topological planar graph G is face-k−colorable if and only if it has a nowhere-zero Z_k-flow.*

Proof. There is a correspondence between the flows on G and the potential differences in the geometric dual G^* (see Kirchhoff's potential law in Section 3.4). This means that in G, each flow can be obtained by assigning a value to each face, and then assigning to each edge the difference between the value of the face on the right and the value of the face on the left. □

We remark that the orientation of G is used only as a reference point for the flows. A flow (or a *circulation* or a *current*) is a fundamental object of graph theory and it appears often in this book. The theorem above led Tutte to study nowhere-zero flows on general graphs. He introduced the *flow polynomial* of a graph G. It again needs an orientation of G to be defined, but does not depend on it.

Theorem 7.2.5. *If A is an Abelian group of order k, then the number of nowhere-zero A-flows on a connected directed graph G is*

$$F(G,k) = \sum_{E'\subset E} (-1)^{|E-E'|} k^{|E'|-r(E')}.$$

Proof. Let T be a spanning tree of G. We recall Observation 2.3.2; the vectors $C(G)_e$, $e \in E(G) \setminus E(T)$, form a cycle basis and the A-flows are exactly the linear combinations of these vectors with the coefficients from A. This has the following consequence. Let $c : E(G)\setminus E(T) \to A$ be an arbitrary mapping. Then there exists exactly one A-flow f of G such that for each edge $e \in E(G) \setminus E(T)$, $f(e) = c(e)$.
Hence for every $E' \subset E$, the number of A-flows of the subgraph (V, E') is $k^{|E'|-r(E')}$. The theorem now follows by the principle of inclusion and exclusion. □

The function $F(G, x)$ is called the *flow polynomial* of G. It is, in a sense, dual to the chromatic polynomial. This is expressed explicitly as follows.

Theorem 7.2.6. *If G is a connected planar topological graph and G^* its dual, then $xF(G, x) = M(G^*, x)$.*

Definition 7.2.7. A *nowhere-zero $k-$flow* is an integer flow whose absolute values are among $1, \cdots, k-1$.

Tutte proved also the following theorem; for a proof see [J1].

Theorem 7.2.8. *A graph has a nowhere-zero $k-$flow if and only if it has a nowhere-zero A-flow for every additive group A of order k.*

Theorem 7.2.8 and Theorem 7.2.4 imply that the Four Color Theorem 2.10.3 is equivalent to

Corollary 7.2.9. *Every orientation of a $2-$edge-connected planar graph has an integer flow with all edge values among $1, -1, 2, -2, 3, -3$.*

Nowhere-zero flows unify three sets of conjectures which live around the Four Color theorem and which have inspired some major research in graph theory. Tutte made the $4-$flow conjecture and the $5-$flow conjecture:

Conjecture 7.2.10. *Every orientation of a $2-$edge-connected graph with no subgraph contractible to the Petersen graph (see Figure 7.1), has a nowhere-zero $4-$flow.*

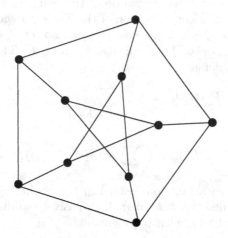

Figure 7.1. The Petersen graph

Conjecture 7.2.11. *Every $2-$edge-connected graph has a nowhere-zero $5-$flow.*

The existence of a $6-$flow is a theorem of Seymour (see [S]).

Theorem 7.2.12. *Every* 2−*edge-connected graph has a nowhere-zero* 6−*flow.*

A *cycle double cover* of a graph G is a family of cycles of G such that every edge appears in exactly two cycles of this family. The following cycle double cover conjecture was asked independently by several researchers; Bondy and Seymour made it particularly popular.

Conjecture 7.2.13. *Every* 2−*edge-connected graph has a cycle double cover.*

A cycle double cover is called *directed* if one can prescribe one of two possible orientations to each of its cycles so that each edge of the graph appears oriented in two opposite ways. Jaeger proposed the directed cycle double cover conjecture:

Conjecture 7.2.14. *Every* 2−*edge-connected graph has a directed cycle double cover.*

Finally there is Fulkerson's conjecture:

Conjecture 7.2.15. *Every* 3−*regular* 2−*edge-connected graph has six perfect matchings such that each edge appears in exactly two of them.*

7.3 Potts, dichromate and ice

In this section we introduce the partition functions of the statistical physics models that correspond to the graph polynomials of the previous section. We start with a definition of a generalization of the Ising model, the *Potts model*. Its partition function is a generalisation of the Tutte polynomial.

Let $G = (V, E)$ be a graph, k positive integer and let $J_e = J_{uv}$ be a weight (coupling constant) associated with an edge $e = uv \in E$. The partition function of the *Potts model* is defined by

$$P^k(G, J_e) = \sum_{s:V \to \{0, \cdots, k-1\}} e^{z(P^k)(s)},$$

where

$$z(P^k)(s) = \sum_{uv \in E} J_{uv} \delta(s(u), s(v)),$$

and δ is Kronecker delta function defined in the beginning of the book. The following function called the *dichromate* is extensively studied in combinatorics. It is clearly equivalent to the Tutte polynomial.

$$B(G, x, y) = \sum_{A \subset E} x^{|A|} y^{c(A)}.$$

We may write

$$P^k(G, J_e) = \sum_{s:V \to \{0, \cdots, k-1\}} \prod_{uv \in E} (1 + z_{uv} \delta(s(u), s(v))) = \sum_{E' \subset E} k^{c(E')} \prod_{uv \in E'} z_{uv},$$

where $z_{uv} = e^{J_{uv}} - 1$. This formula explains why the function $P^k(G, J_e)$ is sometimes called the *multivariate Tutte polynomial*. If J_{uv} are all equal to x we get an expression of the Potts partition function in the form of the dichromate:

$$P^k(G, x) = \sum_s \prod_{uv \in E} e^{x\delta(s(u), s(v))} = \sum_{E' \subset E} k^{c(E')}(e^x - 1)^{|E'|} = B(G, e^x - 1, k).$$

Fortuin and Kasteleyn constructed an extension of the Potts model, called the *random cluster model*. It is a probability space on all spanning subgraphs of G. The partition function of the Fortuin-Kasteleyn model is

$$FK(G, p, q) = \sum_{E' \subset E} p^{|E'|}(1 - p)^{|E - E'|}q^{c(E')},$$

and the probability of $E' \subset E$ is

$$\frac{p^{|E'|}(1 - p)^{|E - E'|}q^{c(E')}}{FK(G, p, q)}.$$

The random cluster model appears to be close to the usual $G(n, p)$ random graphs model (see section 2.8), but is much harder to analyze; the bias is by $q^{c(E')}$: if q is large, the model favours graphs with many components. The following observation is straighforward.

Observation 7.3.1.

$$FK(G, p, q) = (1 - p)^{|E|}B(G, v, q),$$

where $v = p/(1 - p)$.

Ice model. There exist in the nature a number of crystals with hydrogen bonding. The most familiar example is ice. In the 2-dimensional ice model the oxygen atoms fill the vertex set of a planar square lattice. Each vertex has degree 4, and between each adjacent pair of oxygen atoms there is an hydrogen ion. Each ion is located near one or the other end of the edge in which it lies. In 1941, Slater proposed, on the basis of local electric neutrality, that the ions should satisfy the *ice rule*: Of the four ions surrounding each oxygen atom, two are close to it, and two are remote, on their respective bonds. The partition function then equals

$$I(\beta) = \sum_A e^{-\mathcal{E}(A)\beta},$$

where the sum is over all arrangements A of the hydrogen ions that are allowed by the ice rule, and $\mathcal{E}$ denotes the energy of the arrangement.

The arrangements of the ions may be naturally represented by the orientations of edges; the edge-orientation points towards the end of the edge occupied by the ion. In this way each arrangement of the ions corresponds to an orientation of the underlying square lattice; the ice rule is equivalent to saying that at

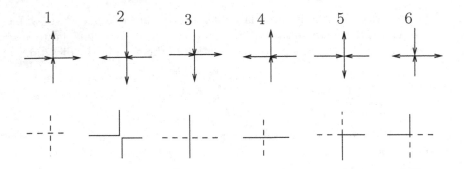

Figure 7.2. Six admissible configurations

each vertex there must be two edges directed towards it, and two edges directed out of it. This leaves us with six ways of arranging the arrows at each vertex, and so the ice model is also called the *six-vertex model* (see Figure 7.2). In a general *ice-type model*, each of these six configurations will have a different energy: let us denote these energies by $\epsilon_1, \cdots, \epsilon_6$. Then the energy of the whole arrangement is $\mathcal{E}(A) = n_1\epsilon_1 + \cdots + n_6\epsilon_6$, where n_i is the number of vertices where configuration i appears. For instances, the ice problem itself is obtained by taking all $\epsilon_i = 0$.

Lieb pointed out that the partition function of the ice model on the toroidal square lattice is equivalent to counting the number of ways of properly coloring the faces of this square lattice with three colours. To see this, consider such a coloring by colors $1, 2, 3$. Place arrows to the edges according to the following rule: If an observer in one face, with color c, looks across an edge to a neigh-

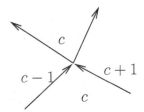

Figure 7.3. Ice and 3-face-colorings

boring face which has color $c + 1$ (modulo 3), then place the arrow pointing to the observer's left, otherwise to his right (see Figure 7.3).

The six-vertex model may also be considered for the planar square grids with some boundary conditions. The boundary conditions mean here that there may be vertices of degree 1 attached to different vertices of the boundary of the square grid, and their incident edges have fixed orientations.

In particular, the states of the six-vertex model of an $n \times n$ square grid with boundary edges pointing inward at both vertical sides and outward at the top and bottom horisontal sides of the grid correspond to the *alternating sign matrices*. An alternating sign matrix (ASM) is a matrix whose elements are $0, 1, -1$ and the non-zero elements in each row and column alternate between $1, -1$ and begin and end with 1. The bijection between the ASMs and the ice configurations is explained by Figure 7.4. The partition function of the ice problem on

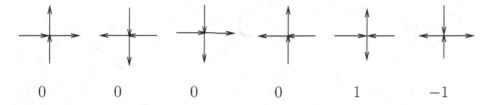

Figure 7.4. Ice and ASM

the square grid was calculated asymptotically by Lieb. There are many fascinating objects and relations to be discovered around the alternating sign matrices. Many of them can be found in the survey articles [P], [GJ]. The exact formula for the number of the ASMs was proved by Zeilberger ([Z]).

Theorem 7.3.2. *The number of $n \times n$ alternating sign matrices is*

$$\frac{1! 4! 7! \cdots (3n - 2)!}{n!(n + 1)!(n + 2)! \cdots (2n - 1)!}.$$

7.4 Graph polynomials for embedded graphs

Embeddings enrich graph polynomials. Let us present two examples. In the first example, we show how the flow polynomial $F(G, x)$ for cubic plane maps may be expressed as a *rotation polynomial* (see section 5.3 for basic facts about the rotation).

Let $G = (V, E)$ be a planar cubic map. If H is an even set of edges in G then it is a vertex-disjoint union of cycles. Each cycle has two possible orientations, clockwise or anti-clockwise. A cycle along with one of the two orientations will be called an *oriented cycle*, and an even set, in which each cycle has a fixed orientation attached, will be called an *oriented even set*. The set of all oriented even sets of edges will be denoted by $\mathcal{OC}$.

Definition 7.4.1. The *rotation polynomial* $R(G, x)$ is defined as follows.

$$R(G, x) = \sum_{C \in \mathcal{OC}} [G, C] x^{2\operatorname{rot}(C)},$$

where $[G, C] = \prod_{v \in V} [v, G, C]$ and the weights $[v, G, C]$ are defined by Figure 7.5, where the edges not in C are dotted. Finally we recall from Section 5.3

that, since C consists of vertex-disjoint cycles, rot(C) is equal to the number of clockwise directed cycles minus the number of anti-clockwise directed cycles of C.

Figure 7.5. Rotation polynomial weights

The proof of the following theorem may be found in [J2].

Theorem 7.4.2. *For every planar cubic map* G

$$R(G, x) = F(G, (x + x^{-1})^2).$$

Arrow coverings. In the second example we show how to express the Potts partition function of a planar graph as an ice partition function.

Let $G = (V, E)$ be an undirected graph, possibly with loops and multiple edges. Let us recall that G together with a fixed cyclic order of the edges incident to each vertex is called a *map*, or a *fatgraph*; see Figure 5.2. We recall that a fatgraph F may be represented by making vertices into *islands* and connecting them by fattened edges (*bridges*) as prescribed by the cyclic orders. This defines a 2-dimensional surface with boundary, which will also be denoted by F. The *genus* $g(F)$ of a fatgraph F is defined as the genus of this surface. It will always be clear from the context whether by F we mean a fatgraph or a surface. For a fatgraph F we will usually denote its underlying graph by $G = G(F)$. Let $V(F)$ be its set of vertices, $E(F)$ its set of edges. We denote the number of connected components of F by $c(F)$, and the number of connected components of the boundary of the surface F by $p(F)$. Further let $v(F) = |V(F)|$, $e(F) = |E(F)|$, $r(F) = |V| - c(F)$ and $n(F) = e(F) - r(F)$. The functions v, e, r, n, c are used for graphs as well. We recall Euler's formula

$$v(F) - e(F) + p(F) = 2(c(F) - g(F)).$$

We further recall that given a fatgraph F, we construct its *medial graph* $M(F)$ as follows (see Figure 5.4): for each bridge, we 'cross' (twist) its boundaries. The vertices of $M(F)$ are these crossings, and the edges are the connections left from the fatgraph F. We call the 'squeezed faces' of $M(F)$ *discs* and we color them black, while the rest of the plane is white. This defines a checkerboard coloring of the plane.

We denote by $S(F)$ the set of the spanning fatsubgraphs of F.

Lemma 7.4.3. *We can express the partition function* $P^k(F, J_e)$ *of the Potts model on the fatgraph* F *as*

$$\sum_{H \in S(F)} k^{c(H)} \prod_{uv \in E(H)} z_{uv} = \sum_{H \in S(F)} k^{1/2(v(H)+p(H)+2g(H))} \prod_{uv \in E(H)} k^{-1/2} z_{uv},$$

(7.1)

where $z_{uv} = e^{J_{uv}} - 1$.

Proof. We describe what a *state* is: we can 'split' each vertex of $M(F)$ so that the black discs incident with it are joined into one, or they are disconnected. There are $2^{e(F)}$ ways to split all the vertices of $M(F)$: these ways are called *states*. We can naturally associate an element of $S(F)$ to each state, and so we identify the set of the states with $S(F)$.

Using $r(F) = v(F) - c(F)$, $n(F) = e(F) - r(F)$ and $g(F) = 1/2(c(F) - p(F) + n(F))$, the lemma follows. $\square$

Next we change the expression of Lemma 7.4.3 into an ice partition function by the technique called *arrow coverings*.

Let us define x by the equation $k^{1/2} = x^{2\pi} + x^{-2\pi}$. Each state $H \in S(F)$ has $p(H)$ faces, which are disjoint polygons (see Figure 5.4). We will think of them as being made up from all the edges of $M(F)$. Let us direct the edges so that each face-polygon is directed, i.e. each polygon corner has one directed edge entering and one directed edge leaving. These orientations are called *allowed arrow coverings*. We can further think of the polygon corners as *transitions* from one directed edge to the other. Let us give to each corner weight x^γ, where γ is the angle to the left through which an observer moving in the direction of the arrows turns when passing through the corner (see Figure 7.6). Since the edges do not overlap, $-\pi < \gamma < \pi$. Still considering a particular state and its allowed

$$\gamma > 0 \qquad\qquad \gamma < 0$$

Figure 7.6. Angles at a corner

arrow coverings, we form the product, over all polygon corners, of these weights

x^γ. Next comes a crucial observation:

The sum of this combined weight over all allowed arrow coverings of a particular state is $k^{p/2}$.

Indeed, walking around a polygon, the observer turns through the total angle 2π if going anti-clockwise, or -2π if going clockwise. These possibilities can occur independently for each polygon and so the total contribution of the polygon is $z^{2\pi} + z^{-2\pi} = k^{1/2}$.

Finally we observe that for each state, the allowed arrow coverings correspond to the admissible orientations of the ice problem on the medial graph $M(F)$. The correspondence is depicted in Figure 7.7, where the first row describes a state with the disconnected black faces and the second row describes a state with the joined black faces. Each of the last two columns of the picture corresponds to a single admissible orientation. Hence we can write the Potts partition function

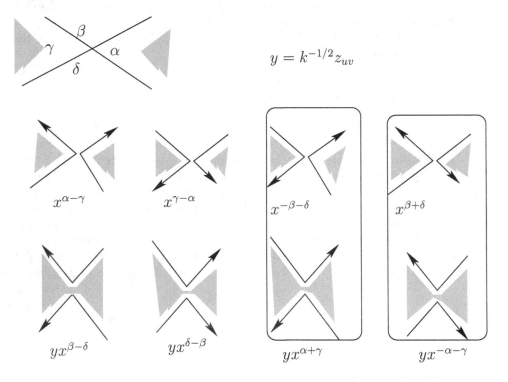

Figure 7.7. Admissible orientations, allowed arrow coverings, and transition weights

$P^k(F, J_e)$ for the *planar fatgraphs* as the partition function of an ice-type model:

$$P^k(F, J_e) = k^{1/2(v(F))} \sum_{D} \prod_{v \in V(F)} w(v, D) \prod_{uv \in E(H)} k^{-1/2} z_{ij}, \qquad (7.2)$$

where the sum is over all admissible orientations of $M(F)$, and $w(v, D)$ contains the information about the weights x^γ and is defined according to Figure 7.7.

This derivation was first discovered for the square-lattice Potts model by Temperley and Lieb. They worked with transfer matrices and discovered that their operators form an interesting algebra, known nowadays as the *Temperley-Lieb algebra*. The Temperley-Lieb algebra with parameter α is the associative algeba generated by $1, e_0, e_1, \ldots, e_n$ and the relations $e_i^2 = \alpha e_i, e_i e_{i+1} e_i = e_i, e_i e_{i-1} e_i = e_i$, and $e_i e_j = e_j e_i$ for $|i - j| \geq 2$.

7.5 Some generalizations

Let us start with the following *q-chromatic function* on graphs:

Definition 7.5.1. Let $G = (V, E)$ be a graph. Let $V = \{1, \cdots, n\}$ and for $k \in \{1, 2, \cdots\}$ let $V(G, k)$ denote the set of all vertex colorings $s : V \to \{0, \cdots, k-1\}$ such that $s(u) \neq s(v)$ whenever $uv \in E$.

$$M_q(G, k) = \sum_{s \in V(G,k)} q^{\sum_{v \in V} s(v)}.$$

Note that $M_q(G, k)|_{q=1}$ is the classical chromatic polynomial of G. As the first example we calculate the $q-$chromatic function of a complete graph. We need some notation first. For a positive integer k, let $(k)_q = q^{k-1} + \cdots + q + 1$ denote a *q-integer*, with the convention that $(0)_q = 0$, and let $(k)!_q = \prod_{1 \leq n \leq k} (n)_q$, with the convention that $(0)!_q = 1$. For $0 \leq n \leq k$ the *q-binomial coefficients* are defined by

$$\binom{n}{k}_q = \frac{(n)!_q}{(k)!_q (n-k)!_q}.$$

These are also known as Gaußian binomial coefficients. A simple q-binomial formula leads to a formula for the summation of the products of distinct powers. This gives the $q-$chromatic function for the complete graph.

Lemma 7.5.2.

$$(a - z)(a - qz) \cdots (a - q^{n-1}z) = \sum_{i=0}^{n} (-1)^i \binom{n}{i}_q q^{i(i-1)/2} a^{n-i} z^i.$$

Proof. We proceed by induction on n. It is easy to check the case $n = 1$. In the induction step assume the statement holds for n and we want to prove it for $n + 1$. Let $y = qz$. We have

$$(a - z)(a - qz) \ldots (a - q^n z) = (a - z)(a - y) \ldots (a - q^{n-1}y) =$$

$$(a - z)[\sum_{i=0}^{n} (-1)^i \binom{n}{i}_q q^{i(i-1)/2} a^{n-i} y^i] =$$

$$\sum_{i=0}^{n} (-1)^i \binom{n}{i}_q q^{i(i-1)/2} a^{n+1-i} z^i q^i + \sum_{i=0}^{n} (-1)^{i+1} \binom{n}{i}_q q^{i(i-1)/2} a^{n-i} z^{i+1} q^i =$$

$$\sum_{i=0}^{n}(-1)^i\binom{n}{i}_q q^{i(i-1)/2}a^{n+1-i}z^i q^i + \sum_{i=1}^{n+1}(-1)^i\binom{n}{i-1}_q q^{i(i-1)/2}a^{n+1-i}z^i =$$

$$\binom{n}{0}_q a^{n+1} + (-1)^{n+1}\binom{n}{n}_q q^{n(n+1)/2}z^{n+1}+$$

$$\sum_{i=1}^{n}(-1)^i q^{i(i-1)/2}a^{n+1-i}z^i[q^i\binom{n}{i}_q + \binom{n}{i-1}_q] =$$

$$\sum_{i=0}^{n+1}(-1)^i\binom{n+1}{i}_q q^{i(i-1)/2}a^{n+1-i}z^i$$

since it may be observed directly that

$$q^i\binom{n}{i}_q + \binom{n}{i-1}_q = \binom{n+1}{i}_q.$$

$\square$

Corollary 7.5.3. *For a positive integer k, the q-chromatic function of the complete graph on $n \leq k$ vertices is given by*

$$M_q(K_n, k) = n!\binom{k}{n}_q q^{n(n-1)/2}$$

and $M_q(K_n, k) = 0$ for $n > k$.

The next theorem provides a natural way to extend the q–chromatic function from positive integers to real numbers, by extending q–integers $(k)_q$ to q-*numbers* $(y)_q = \frac{q^y-1}{q-1}$ for real variables y and $q \neq 1$ (and $(y)_1 = y$ by continuity, $\lim_{q\to1}(y)_q = y$.) If $G = (V, E)$ is a graph and $A \subset E$ then let $C(A)$ denote the set of the connected components of graph (V, A) and $c(A) = |C(A)|$. If $W \in C(A)$ then let $|W|$ denote the number of vertices of W.

Theorem 7.5.4. *For positive integers k,*

$$M_q(G, k) = \sum_{A\subseteq E}(-1)^{|A|}\prod_{W\in C(A)}(k)_{q^{|W|}}.$$

Proof. We connect the principle of inclusion and exclusion (PIE) with the geometric series formula:

$$M_q(G, k) = \sum_{s:V\to\{0,\cdots,k-1\}} q^{\sum_{v\in V} s(v)} - \sum_{s\in\bigcup_{e\in E} I_e} q^{\sum_{v\in V} s(v)},$$

where $I_e, e = uv \in E$, denotes the set of functions $s : V \to \{0, \cdots, k-1\}$ for which $s(u) = s(v)$.
By PIE this equals

$$\sum_{A\subseteq E}(-1)^{|A|}\sum_{s\in I_A} q^{\sum_v s(v)} =$$

$$\sum_{A \subseteq E} (-1)^{|A|} \prod_{W \in C(A)} \sum_{0 \leq i \leq k-1} q^{i|W|} = \sum_{A \subseteq E} (-1)^{|A|} \prod_{W \in C(A)} (k)_{q^{|W|}}.$$

$\square$

The formula of Theorem 7.5.4 leads naturally to the definition of the $q-dichromate$.

Definition 7.5.5. For variables x, y

$$B_q(G, x, y) = \sum_{A \subseteq E} x^{|A|} \prod_{W \in C(A)} (y)_{q^{|W|}}.$$

Note that $B_{q=1}(G, x, y) = B(G, x, y)$ and $M_q(G, k) = B_q(G, -1, k)$ by Theorem 7.5.4.
Let $x_1, x_2, \cdots$ be commuting indeterminates and let $G = (V, E)$ be a graph. The $q-$chromatic function $M_q(G, y)$ restricted to non-negative integers y is the principal specialization of X_G, the *symmetric function generalization of the chromatic polynomial* defined by Stanley as follows:

Definition 7.5.6.

$$X_G(x_0, x_1, \cdots) = \sum_{s \in \cup_{k \in \mathbb{N}} V(G, k)} \prod_{v \in V} x_{s(v)},$$

where the sum is over all proper colorings of G by $\{0, 1, \cdots\}$.

Therefore $M_q(G, k) = X_G(x_i = q^i (0 \leq i \leq k - 1), x_i = 0 (i \geq k))$. Stanley further defined the *symmetric function generalization of the bad coloring polynomial*:

Definition 7.5.7.

$$XB_G(t, x_0, x_1, \cdots) = \sum_{s: V \to \{0, 1, \cdots\}} (1 + t)^{b(s)} \prod_{v \in V} x_{s(v)},$$

where the sum ranges over ALL colorings of G by $\{0, 1, \cdots\}$ and $b(s) := |\{uv \in E : s(u) = s(v)\}|$ denotes the number of monochromatic edges of f.

Noble and Welsh defined the *U-polynomial* and showed that it is equivalent to XB_G. I. Sarmiento proved that the *polychromate* defined by Brylawski is also equivalent to the *U*-polynomial.

Definition 7.5.8.

$$U_G(z, x_1, x_2 \cdots) = \sum_{A \subseteq E(G)} x(\tau_A)(z - 1)^{|A| - |V| + c(A)},$$

where $\tau_A = (n_1 \geq n_2 \geq \cdots \geq n_l)$ is the partition of $|V|$ determined by the connected components of A, $x(\tau_A) = x_{n_1} \cdots x_{n_l}$.

The motivation for the work of Noble and Welsh was a series of papers by Chmutov, Duzhin and Lando where they observed that the U-polynomial, evaluated at $z = 0$ and applied to the intersection graphs of chord diagrams, satisfies the $4T$−relation of the weight systems (see Section 8.10). Hence the same is true for $M_q(G, k)$ for each positive integers k since it is a particular evaluation of the U-polynomial.

Observation 7.5.9. *For a positive integer k,*

$$M_q(G, k) = (-1)^{|V|} U_G(0, x_1, x_2, \cdots)|_{x_i = -(k)_{q^i}}.$$

On the other hand, it seems plausible that the q−dichromate determines the U-polynomial. If true, then the q−dichromate would provide a compact representation of all the multivariate generalizations of the Tutte polynomial mentioned above.

The next theorem states how the q−dichromate is related to the partition function of the Potts model (with a variable external field).

Theorem 7.5.10.

$$\sum_{A \subseteq E} \prod_{W \in C(A)} (k)_{q^{|W|}} \prod_{uv \in A} z_{uv} = \sum_{s} q^{\sum_{v \in V} s(v)} e^{E(P^k)(s)},$$

where $z_{uv} = e^{J_{uv}} - 1$ as above.

Proof. Let $W(A, k)$ denote the set of all colorings $s : V \to \{0, \cdots, k-1\}$ whose set of monochromatic edges contains A.
We have

$$P_q^k(G, J_e) = \sum_{s} q^{\sum_{v \in V} s(v)} \prod_{uv \in E} (1 + z_{uv} \delta(s(u), s(v))) =$$

$$\sum_{s} q^{\sum_{v \in V} s(v)} \sum_{A \subseteq E} \prod_{uv \in A} z_{uv} \delta(s(u), s(v)) =$$

$$\sum_{A \subseteq E} \sum_{s \in W(A,k)} q^{\sum_{v \in V} s(v)} \prod_{uv \in A} z_{uv} =$$

$$\sum_{A \subseteq E} \prod_{W \in C(A)} (k)_{q^{|W|}} \prod_{uv \in A} z_{uv}.$$

$\square$

7.6 Tutte polynomial of a matroid

The Tutte polynomial of a matroid is defined as follows.

Definition 7.6.1. Let M be a matroid on set E. For $A \subset E$ let $r(A)$ denote the rank of A in M. Then let

$$T(M, x, y) = \sum_{A \subset E} (x - 1)^{r(E) - r(A)} (y - 1)^{|A| - r(A)}.$$

Example 7.6.2. If G is a graph then $T(G, x, y) = T(M(G), x, y)$.

If M is a matroid and M^* is its dual, then by Proposition 4.5.2, $r^*(E) - r^*(A) = |A| - r(A)$ and we immediately get the *duality of the Tutte polynomial:*

$$T(M, x, y) = T(M^*, y, x).$$

If a linear code $\mathcal{C}$ of length n (see section 6.4) is given as the row space of a $k \times n$ matrix A over a field $\mathbb{F}$, i.e. $\mathcal{C} = \{xA; x \in \mathbb{F}^k\}$, then we will denote it by $\mathcal{C} = C(\mathbb{F}, A)$. We recall that $C(\mathbb{F}, A)^* = \{x \in \mathbb{F}^n; Ax = 0\}$.
We denote by $M(A)$ the matroid represented by the columns of A. If $\mathcal{C} = C(\mathbb{F}, A)$ and a matroid M is represented by the columns of A, then we say that $\mathcal{C}$ is the *cut space* of M and $\mathcal{C}^*$ is the *cycle space* of M. This coincides with the notions introduced for graphs in Section 2.3.
Greene proved that the weight enumerator of a linear code over a finite field $\mathbb{F} = GF(q)$ is an evaluation of the Tutte polynomial of a matroid represented by the generating matrix of the code. His theorem reads as follows.

Theorem 7.6.3. *Let C be a linear code of length n and dimension k over $GF(q)$, which is the cycle space of a matroid M. Then for $0 \neq t \neq 1$,*

$$A_C(t) = (1 - t)^k t^{n-k} T\left(M(GF(q), C), \frac{1 + (q-1)t}{(1-t)}, \frac{1}{t}\right).$$

Chapter 8

Knots

A *knot* is a subset of $\mathbb{R}^3$ which is homeomorphic to a circle. We restrict ourselves here to *tame knots*, i.e. those which are simple closed polygons in $\mathbb{R}^3$. A *link* with k components is a subset of $\mathbb{R}^3$ which consists of k disjoint knots (components of the link). A link is *oriented* if each component is prescribed one of the two possible directions of traversal. Since the beginning of the knot theory, knots are regularly projected onto the plane and thus represented by a *(planar knot) diagram*. The notion of the knot or link diagram is very intuitive. Some examples of oriented diagrams appear in Figure 8.1. A link diagram is simply a 4-regular topological planar graph where the vertices are replaced by the *crossings*. At each crossing it is specified which transition goes up and which one goes down. For oriented links this leads to two kinds of crossings, denoted by the

Figure 8.1. Unknot, unknot, right-handed trefoil, figure eight knot

signs $+$ and $-$. These signs are defined by Figure 8.2. Two knots are equivalent if they can be transformed into each other by a continuous deformation of the ambient space. The equivalence may be captured combinatorially by the Δ-move in $\mathbb{R}^3$ and its reverse move. If K is a knot in $\mathbb{R}^3$, i.e. a closed simple polygon by our assumption, then a Δ-move consists in replacing a straight line segment l of K by the other two sides of a triangle T having sides l, k, j. It is assumed that T and K intersect only in l.

An *unknot* is any knot which is equivalent to the knot whose diagram is a circle. A natural task of listing all the knots leads to the problem of deciding whether two knots represented by their diagrams are equivalent. This is extensively

Figure 8.2. Oriented signs

studied and is considered to be a hard problem. The classification problem for knots also led to study *knot invariants*. Knot invariants are functions on the knot diagrams which give the same answer for the diagrams of equivalent knots. It is a seminal open problem to find a knot invariant that assigns a polynomial to every diagram (a *polynomial knot invariant*) and distinguishes every two inequivalent knots.

8.1 Reidemeister moves

An initial step in understanding knot invariants was a description of a system of basic moves for the diagrams, which capture the equivalence of the knots in $\mathbb{R}^3$ given by the Δ-moves. These are the *Reidemeister moves*. The moves are described in Figure 8.3 for the unoriented diagrams. The Reidemeister moves for

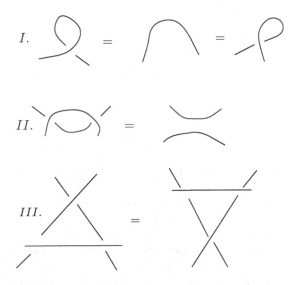

Figure 8.3. Three types of Reidemeister moves for unoriented knots

the oriented diagrams may be grouped into three types following the unoriented case. With regard to the orientation, there are clearly 4 types of the oriented I-move. It is not hard to observe that all the oriented II-moves together with all the oriented III-moves can be generated by the moves II(A), II(B), III(+) described in Figure 8.4. Here comes the theorem of Reidemeister.

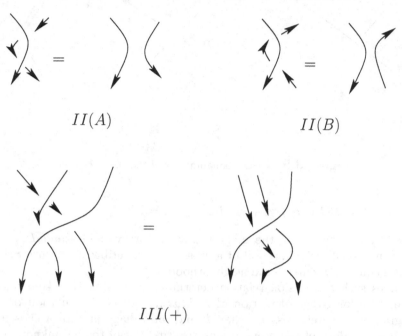

$$II(A) \qquad\qquad II(B)$$

$$III(+)$$

Figure 8.4. Three types of Reidemeister moves for oriented knots

Theorem 8.1.1. *Two knots are equivalent if and only if any diagram of one can be obtained from any diagram of the other by some finite sequence of Reidemeister moves.*

Two diagrams that can be obtained from each other by II-moves and III-moves only are said to be *regular isotopic*.

Example 8.1.2. The *writhe* is defined by $\omega(\mathcal{K}) = n_+ - n_-$, where n_+ is the number of positive crossings of the oriented link diagram $\mathcal{K}$ and n_- is the number of negative crossings of $\mathcal{K}$. For knots, this does not depend on the chosen orientation. It is clear that $\omega(\mathcal{K})$ is regular isotopy invariant and that, under type-I move, it changes by one.

8.2 Skein relation

Skein relations are recursion relations for knot invariants which describe the value of the invariant on a link diagram L as a simple function of values of the invariant on the diagrams obtained by a local change of L. The skein relation often describes the knot invariant in a most elementary way. Let us consider oriented knots first. A basic skein relation is

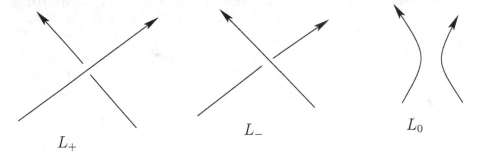

L_+ L_- L_0

Figure 8.5. Three configurations of the skein relation

$$xP(L_+, x, y, z) + yP(L_-, x, y, z) = zP(L_0, x, y, z),$$

where P is a function on the diagrams, x, y, z are variables and L_+, L_-, L_0 behave in a small neighborhood of a crossing as described in Figure 8.5, and they are equal to L outside the neighborhood.

Why does such a recursion relation determine P? Using the skein relation, we can transform the calculation of $P(L, x, y, z)$ to the calculation of P on a diagram with fewer crossings than L and on a diagram with a chosen sign of L altered. Both of these operations eventually lead to the unknot. Hence $P(L, x, y, z)$ equals a linear combination of terms of the form $P(\text{unknot}, x, y, z)$ times a polynomial in the variables $x, x^{-1}, y, y^{-1}, z, z^{-1}$. Usually one specifies $P(\text{unknot}, x, y, z) = 1$. The real problem is to show the invariance under the Reidemeister moves. The above skein relation defines a three variable knot invariant P, called the *Homfly polynomial.*

8.3   The knot complement

Two equivalent links have homeomorphic complements. The opposite implication is not true. For instance, a link and its mirror image have homeomorphic complements but need not be equivalent. The mirror image is defined from a diagram by exchanging the signs of all the crossings. It was proved by Gordon and Luecke that two knots with homeomorphic complements are equivalent under Reidemeister moves and taking mirror images.

The *knot group* is the fundamental group $\pi_1(\mathbb{R}^3 - \mathcal{K})$ of the knot complement. There is an easy way, called the *Wirtinger presentation*, to describe the knot group. Let K be a directed knot diagram. An *arc* is a segment of K which starts at an undercrossing and continues until the next undercrossing. A diagram with r crossings clearly has r arcs. Let u be a point lying 'above' the oriented knot

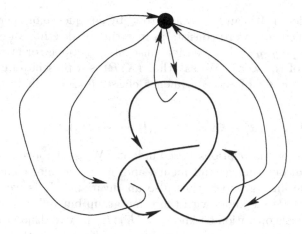

Figure 8.6. Loops around every arc

diagram. For any arc a of the diagram, we consider an oriented loop l_a in $\mathbb{R}^3$ which starts at u and goes around a in positive direction (see Figure 8.6). Let r be the number of crossings of the diagram, which is also its number of arcs. The r loops $l_1, \cdots, l_r$ represent homotopy classes in $\pi_1(\mathbb{R}^3 - \mathcal{K})$, and we have r relations $s_1, \cdots, s_r$, one at every crossing, described in Figure 8.7. The next

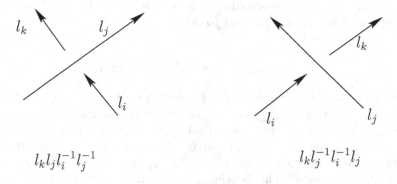

Figure 8.7. Relations in the Wirtinger presentation

theorem of van Kampen asserts that these are sufficient to give a presentation of the knot group $\pi_1(\mathbb{R}^3 - \mathcal{K})$. Let us briefly recall that $(G, \cdot)$ is a group if G is a non-empty set and $\cdot$ is a binary operation on G which is associative, there is unity $1 \in G$ satisfying $1 \cdot g = g \cdot 1 = g$ for each $g \in G$; and for each $g \in G$ there is $g^{-1} \in G$ so that $g \cdot g^1 = g^{-1} \cdot g = 1$. A *presentation* of the group G is a pair (X, R) where X is a set and R is a set of non-empty words with letters from X, which satisfies the following property:

Let X^* denote the set of the words with letters from X. Two words W, Q from

X^* are equivalent if W can be reduced to Q by a sequence of operations. Each operation applied to a word consists in inserting or deleting an element of R, or a word of type $g \cdot g^{-1}$, between its consecutive symbols, or at the beginning or at the end of the word. We say that (X, R) is a presentation of G if G is isomorphic to the set of the equivalence classes of X^* with the operation of concatenation.

Theorem 8.3.1. $\pi_1(\mathbb{R}^3 - \mathcal{K}) = (l_1, \cdots, l_r; s_1, \cdots, s_r)$.

In fact, one of the relations is redundant. We saw above that equivalent knots have homeomorphic complements, and hence the knot complement is an invariant. The knot group is thus also an invariant. It is weaker than the complement: there are knots with the same group but different complements. The *sum* is a basic operation on knots: let K_1, K_2 be two disjoint knots and let l_1, l_2 be line segments in K_1, K_2. Then the sum $K_1 + K_2$ is the knot which we get from K_1 by exchanging l_1 by $K_2 \setminus l_2$. A knot is *prime* if it is not the unknot, and cannot be expressed as the sum of two non-trivial knots. Two *prime* knots with isomorphic groups do have homeomorphic complements.

8.4   The Alexander-Conway polynomial

The considerations of the knot group led to the discovery of the first *polynomial knot invariant*, the Alexander-Conway polynomial. The Alexander-Conway polynomial $\Delta(L, t)$ was discovered in 1928 by Alexander. In 1970 Conway constructed a knot invariant equivalent to $\Delta(L, t)$.
The skein relation is

$$\Delta(L_+, t) - \Delta(L_-, t) = (t^{-1/2} - t^{1/2})\Delta(L_0, t),$$

and $\Delta(\text{unknot}, t^{-1/2} - t^{1/2}) = 1$. (See Figure 8.5 for L_+, L_-, L_0).
A remarkable fact about the Alexander-Conway polynomial is that it can be defined directly from the circle embedded in $\mathbb{R}^3$. Any presentation of the knot group leads to an $(r-1) \times (r-1)$ matrix whose determinant is the Alexander polynomial $\Delta(t)$, up to a power of t. We explain the construction for the Wirtinger presentation. The Wirtinger presentation leads to the following r by r transition matrix $B_\mathcal{K} = (\beta_{ij})$ called the *Bureau matrix*. We introduce it together with the notion of the *arc-graph* which will be useful in the study of the Jones polynomial as well.
Let $a_1, \cdots, a_r$ be the arcs of $\mathcal{K}$ which we label so that each arc a_i ends at the crossing i. We will single out a specific arc of $\mathcal{K}$ which we decorate by $\star$. Without loss of generality, we may assume that the last arc a_r is decorated by $\star$. We denote by K the *long knot* obtained by removing $\star$ from $\mathcal{K}$. Given $\mathcal{K}$, we construct a weighted directed graph $G_\mathcal{K}$.
The *arc-graph* $G_\mathcal{K}$ has r vertices $1, \ldots, r$, r *blue directed edges* $(v, v+1)$ (v taken modulo r) and r *red directed edges* (u, v), where at the crossing u the arc that crosses over is labeled by a_v.

Each vertex v of $G_\mathcal{K}$ is equipped with a sign $\mathrm{sign}(v)$ equal to the sign of the corresponding crossing v of $\mathcal{K}$. The edges of $G_\mathcal{K}$ are equipped with a weight β, where the weight of the blue edge $(v, v+1)$ is $t^{-\mathrm{sign}(v)}$, and the weight of the red edge (u, v) is $1 - t^{-\mathrm{sign}(u)}$. Here t is a variable. Finally, G_K denotes the digraph obtained by deleting vertex r from $G_\mathcal{K}$.

It is clear that from every vertex of $G_\mathcal{K}$, the blue outdegree is 1, the red outdegree is 1, and the blue indegree is 1. It is also clear that $G_\mathcal{K}$ has a Hamiltonian cycle that consists of all the blue edges. We denote by e_i^b (e_i^r) the blue (red) edge *leaving vertex i.*

Example 8.4.1. For the figure 8 knot we have:

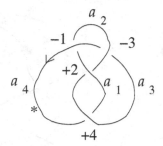

Its arc-graph $G_\mathcal{K}$ with the ordering and signs of its vertices is given by

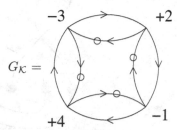

$$G_\mathcal{K} =$$

where the blue edges are the ones with circles on them.

The Bureau matrix $B_\mathcal{K}$ has entries equal to the weights β of the arc-graph. Let B_K denote the submatrix obtained by removing the row and the column indexed by the $*$.

For example for the figure 8 knot we have:

$$B_\mathcal{K} = \begin{bmatrix} 0 & t & 0 & 1-t \\ 1-\bar{t} & 0 & \bar{t} & 0 \\ 0 & 1-t & 0 & t \\ \bar{t} & 0 & 1-\bar{t} & 0 \end{bmatrix}.$$

Theorem 8.4.2. *For every knot diagram $\mathcal{K}$ we have:*

$$\Delta(\mathcal{K}, t) =_t \det(I - B_K),$$

where the equality is up to a power of t.

Since the Alexander polynomial may be expressed by a determinant, it is efficiently computable.

8.5   Braids and the braid group

In this section we describe the *braid group*. The Yang-Baxter equation from
Section 6.8 appears as a defining relation of the braid group. We show how
the Yang-Baxter equation appears very naturally from the third Reidemeister
move.

A *braid* of m strings (m-braid) is constructed as follows: take m distinct points
$u_1, \cdots, u_m$ in a horizontal line and link them to distinct points $v_1, \cdots, v_m$ lying in a parallel line by m disjoint simple strings $l_1, \cdots, l_m$. The strings are
required to 'run downwards', as illustrated in Figure 8.8. The linking deter-

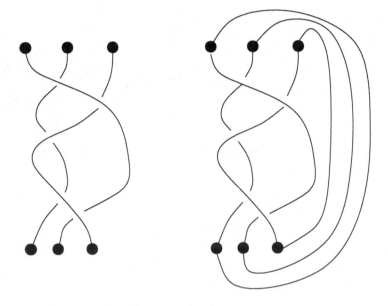

Figure 8.8. A braid and its closure

mines a permutation π of $1, \cdots, m$ so that each u_i is linked by l_i to $v_{\pi(i)}$. The
permutation π is called the *permutation* of the braid. The braid is *closed* by
joining the points u_i, v_i, see Figure 8.8. Alexander proved that every link is
equivalent to the closure of a braid. There is a natural way to compose braids
with the same number of strings. Let a braid b have initial points $u_1, \cdots, u_m$
and terminal points $v_1, \cdots, v_m$, and a braid b' have initial points $u'_1, \cdots, u'_m$
and terminal points $v'_1, \cdots, v'_m$. The composition bb' is obtained by identifying
each pair v_i, u'_i. Under this composition, the classes of the equivalent m-braids
form a group, called the *braid group* $\mathcal{B}_m$. The braid group $\mathcal{B}_m$ is generated by
the *elementary braids* σ_i, σ_i^{-1}, $i = 1, \cdots, m-1$, which represent transpositions
$(i, i+1)$. Each elementary braid has exactly one crossing and thus the ele-
mentary braids come in pairs σ_i, σ_i^{-1} that have the opposite sign of this unique
crossing. Defining relations of the braid group were proved by Artin to be

(1) $\sigma_i\sigma_j\sigma_i = \sigma_j\sigma_i\sigma_j$ if $|i-j|=1$,

(2) $\sigma_i\sigma_j = \sigma_j\sigma_i$ if $|i-j|>1$.

The relation (1) describes the Reidemeister move III(+) (see Figure 8.4). It is the Yang-Baxter equation of Section 6.8.

8.6 Knot invariants and vertex models

In a vertex model we are given a graph $G = (V,E)$. A *state* is defined as an arbitrary function which assigns to each edge an element of $\{1,\ldots,q\}$. We also say that each element is assigned a *color*. The partition function is now defined by

$$I(G) = \sum_{s:E\to\{1,\ldots,q\}} \prod_{v\in V} I(v,s),$$

where $I(v,s)$ is a function of the colors of the edges incident with v.

An important case is when G is directed and each vertex has two incoming and two outgoing incident directed edges. We already saw this as the *ice model* in Section 7.3. In this case, the vertex weights $I(v,s)$ may be described by a square $q^2\times q^2$ matrix where the rows are indexed by the colorings of the two incoming edges and the columns are indexed by the colorings of the two outgoing edges. Such a matrix is known as the R-matrix in knot theory. We further restrict ourselves to $q = 2$. In this case the R-matrices are 4×4 matrices. We will denote by $R(v)^{cd}_{ab}$ the element of the R-matrix which describes the weight of vertex v when the two incoming edges have colors a,b and the two outgoing edges have colors c,d. If a graph G is a knot diagram then the weights of vertices are typically described by two matrices, R^+ and R^-, one for each sign of the crossings (see Figure 8.9). Which $R-$matrices lead to knot invariants? We

$(R^+)^{cd}_{ab}$ $\qquad\qquad$ $(R^-)^{cd}_{ab}$

Figure 8.9. R-matrix correspondence

saw in Section 8.5 that the solutions of the Yang-Baxter equation lead naturally to the invariance of type III Reidemeister moves. Another basic idea in the constructions of quantum knot invariants is that vertex models on *topological graphs* embedded on a 2-dimensional surface can be conveniently endowed with a local geometric information. This seems to have been used first by Baxter, when he formulated the *Potts model* on a planar topological graph as a vertex model; see Section 7.3. Jones and Turaev carried this idea over to knots. They observed that it may be advantageous to enhance the $R-$matrix of each vertex by factors depending on the angles between the incident edges. These factors may differ from vertex to vertex and they may often be regrouped to the global *rotation* contribution. An example is given in the next section, where the Alexander polynomial is described as such an enhanced vertex model.

8.7   Alexander-Conway as a vertex model

We will use the notion of Section 8.6. In particular we consider each knot diagram as a directed plane graph where each vertex is assigned by a sign, and has exactly two incoming and two outgoing incident edges. The $R-$matrices R^+ and $R^- = (R^+)^{-1}$, which describe the vertex weights $I(v, s)$ (see Section 8.6), are given for the Alexander-Conway polynomial by

$$(R^+)^{0,0}_{0,0} = 1 \quad (R^+)^{1,0}_{0,1} = (R^+)^{0,1}_{1,0} = -t \quad (R^+)^{1,1}_{1,1} = -t^2 \quad (R^+)^{0,1}_{0,1} = 1 - t^2$$
$$(R^-)^{0,0}_{0,0} = 1 \quad (R^-)^{1,0}_{0,1} = (R^-)^{0,1}_{1,0} = -t^{-1} \quad (R^-)^{1,1}_{1,1} = -t^{-2} \quad (R^-)^{1,0}_{1,0} = 1 - t^{-2}$$

All other entries of the $R-$matrix are zero.

We say that a 2-coloring (state) s is *admissible* if $\prod_{\text{vcrossing}} I(v, s) \neq 0$ (see Section 8.6). In each admissible coloring, each of the two sets of monochromatic edges defines a collection of closed oriented loops in the plane. Let us denote these sets by s_0 and s_1 according to the color. We now let $\text{rot}(s_0)$ be the rotation of s_0 introduced in Section 5.3. The following theorem was obtained by Jaeger, Kauffman and Saleur (see [JKS]).

Theorem 8.7.1. *Let e be an arbitrary edge of the diagram $\mathcal{K}$. Then*

$$\Delta(\mathcal{K}, t^2) =_t (-t)^{-\omega(\mathcal{K})} \sum_{state\ s; s(e)=1} (-1)^{\text{rot}(s_0)} \prod_{v\ crossing} I(v, s),$$

where the equality is up to a power of t.

8.8   The Kauffman derivation of the Jones polynomial

Spectacular advances in knot theory based on the connections with theoretical physics were initiated by Jones' discovery of the Jones polynomial $V(t)$ in 1985. In 1986 Kauffman produced a beautiful derivation of the Jones polynomial using the bracket polynomial, defined in a statistical mechanical way as a state sum. The following skein relation was obtained by Jones in the original paper:

$$t^2 V(L_+, t) - t^{-2} V(L_-, t) = (t - t^{-1}) V(L_0, t),$$

and $V(\text{unknot}, t) = t + t^{-1}$.

We will also consider the normalized version of the Jones polynomial defined by
$J(\mathcal{K})(t) = V(\mathcal{K})(t^{1/2})/V(\text{unknot})(t^{1/2})$.

The Jones polynomial was originally understood in terms of representations of quantum groups, and Witten gave a quantum field theory interpretation of the Jones polynomial as the expectation value of Wilson loops of a 3-dimensional theory, the Cherns-Simons theory. This connection between statistical physics, knot theory, quantum field theory and combinatorics has kept mathematicians and physisists busy for decades. The Homfly polynomial was defined shortly after Jones' discovery, and it specializes to both the Alexander and Jones polynomials.

We will now show a construction of Kauffman, which derives the Jones polynomial from the Potts partition function. Let $\mathcal{K}$ be a directed knot diagram, where each crossing c has a *directed sign* $\text{sign}(v)$ associated with it, and two arcs entering and leaving it. We consider $\mathcal{K}$ as a directed plane graph. Given $\mathcal{K}$, we construct its *medial graph* $M = (V(M), E(M))$. The construction is similar as in the study of Gauß codes in Section 5.2. It can be described as follows. First we color the faces of $\mathcal{K}$ white and black so that the neighbouring faces receive a different color, and the outer face is white. Let $b(v)$ be the *undirected sign* of crossing v, induced by this coloring (see Figure 8.10). Let $V(M)$ be the

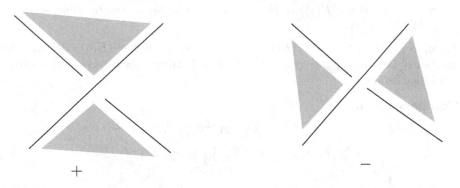

$+$ $-$

Figure 8.10. Undirected sign

set of the black faces, where two vertices are joined by an edge in $E(M)$ if the corresponding faces share a crossing. Note that M is again a plane graph. For the edge e of M let $b(e)$ be the 'undirected' sign of the crossing shared by the end-vertices of e.

Let us now describe what a *state* will be: we can 'split' each crossing v of $\mathcal{K}$ so that the white faces incident with v are merged into one face and the black faces are disconnected, or vice versa. If $\mathcal{K}$ has r crossings then there are 2^r ways to split all of them: these are called *states*. After performing all the splittings of a state s, we are left with a set of disjoint non-self-intersecting cycles in the plane;

let $S(s)$ denote their number. The state s will be identified with a function s from the crossings of $\mathcal{K}$ to $\{-1, 1\}$, where $s(v) = 1$ if the black faces are merged at v and $s(v) = -1$ if the white faces are merged at v. The following theorem was discovered by Kauffman (see [K1]).

Theorem 8.8.1. *Let $\mathcal{K}$ be a directed knot diagram. The following function $f_K(\mathcal{K})$ is a knot invariant:*

$$f_K(\mathcal{K}, A) = (-A)^{-3\omega(\mathcal{K})} \sum_s (-A^2 - A^{-2})^{S(s)-1} A^{\sum_{v \text{ crossing}} b(v)s(v)}.$$

Moreover the Jones polynomial is given by

$$J(\mathcal{K}, A^{-4}) = f_K(\mathcal{K}, A).$$

Each state s determines a subset of edges $E(s) \subset E(M)$ which are not deleted by the splittings of s, and it is easy to see that this gives a natural bijection between the set of the states and the subsets of $E(M)$. Moreover for each state s

$$\sum_{v \text{ crossing}} b(v)s(v) = - \sum_{e \in E(M)} b(e) + 2 \sum_{e \in E(s)} b(e).$$

Proposition 8.8.2.

$$S(s) = 2c(E(s)) + |E(s)| - |V(M)|,$$

where we recall that $c(E(s))$ denotes the number of the connected components of graph $(V(M), E(s))$.

Proof. Note that $S(s) = f(E(s)) + c(E(s)) - 1$, where $f(E(s))$ denotes the number of faces of a plane graph $(V(M), E(s))$. Hence the formula follows from the Euler formula. $\qquad\square$

Corollary 8.8.3.

$$f_K(\mathcal{K}, A) = (-A)^{-3\omega(\mathcal{K})}(-A^2 - A^{-2})^{-|V(M)|-1} A^{-\sum_{e \in E(M)} b(e)} \times$$

$$\sum_{E' \subset E(M)} (-A^2 - A^{-2})^{2c(E')} \prod_{e \in E'} (-A^2 - A^{-2}) A^{2b(e)}.$$

This obviously provides an expression for the Jones polynomial of a knot diagram $\mathcal{K}$ as the Potts partition function of its medial graph M. In particular, if $\mathcal{K}$ is alternating and thus all the signs $b(e)$ are the same, its Jones polynomial can be expressed using the Tutte polynomial. Having the q-dichromate (see Definition 7.5.5) in mind, we may ask:

Question 8.8.4. Is

$$\sum_{E' \subset E(M)} \prod_{e \in E'} (-A^2 - A^{-2}) A^{2b(e)} \prod_{W \in C(E')} ((-A^2 - A^{-2})^2)_{q^{|W|}}$$

times an appropriate prefactor (equal to that for $f_K(\mathcal{K}, A)$ when $q = 1$) also a knot invariant? We recall that $C(E')$ denotes the set of the components of the graph $(V(M), E')$.

8.9 Jones polynomial as vertex model

We will use the notions of Section 8.6. In particular we again consider each knot diagram as a directed plane graph where each vertex is decorated by a sign, and has exactly two incoming and two outgoing incident edges. We recall from Section 8.6 that the partition function of a vertex model may be written as

$$I(G) = \sum_{\text{states}} \prod_{v \in V} I(v, s).$$

For the Jones polynomial, the weight $I(v, s)$ is given, according to Section 8.6, by the R-matrices R^+ and $R^- = (R^+)^{-1}$ below.

$$(R^+)^{0,0}_{0,0} = (R^+)^{1,1}_{1,1} = -q \qquad (R^+)^{1,0}_{0,1} = (R^+)^{0,1}_{1,0} = 1 \qquad (R^+)^{0,1}_{0,1} = q^{-1} - q$$
$$(R^-)^{0,0}_{0,0} = (R^-)^{1,1}_{1,1} = -q^{-1} \qquad (R^-)^{1,0}_{0,1} = (R^-)^{0,1}_{1,0} = 1 \qquad (R^-)^{1,0}_{1,0} = q - q^{-1}$$

and all other entries of the R matrix are zero.

As before, we say that a 2-coloring (a state) s is *admissible* if $\prod_{v\text{crossing}} I(v, s) \neq 0$. In each admissible coloring, each of the two sets of the monochromatic edges defines a collection of closed oriented loops in the plane. Let us denote these sets by s_0 and s_1 according to the color. We now let $\text{rot}(s) = \text{rot}(s_0) - \text{rot}(s_1)$, where $\text{rot}(s_i)$ is the rotation introduced in Section 5.3. Turaev ([TV]) proved the following

Theorem 8.9.1.

$$V(K, q) = (-q^2)^{-\omega(K)} \sum_{\text{2-coloring } s} q^{\text{rot}(s)} \prod_{v \text{ crossing}} I(v, s).$$

We remark that another vertex model expression for the Jones polynomial can be obtained from Theorem 8.8.1, that expresses the Jones polynomial as the Potts partition function, and from Equation 7.2, that expresses the Potts partition function of a planar topological graph as a vertex model.

8.10 Vassiliev invariants and weight systems

An important role within the theory of quantum knot invariants (cousins of the Jones polynomial) has been played by the *finite type invariants (Vassiliev invariants)* and their *weight systems*. We follow [B1] in this brief introduction. We consider a generalization of the knot diagrams where we change some of the crossings into normal vertices. These new vertices are called *double points*. Let $\mathcal{K}^m$ be the set of all *m-singular knot diagrams*, i.e., knot diagrams with m double points.

Let V be an arbitrary invariant of oriented knots. We can extend V to $\mathcal{K}^m$ by repeatedly using the formula of Figure 8.11. We say that V is of type m if its extension to $\mathcal{K}^{m+1}$ vanishes identically. We say that V is of finite type if it is of

$$V(\; \times \;) \; = \; V(\; \times \;) \; - \; V(\; \times \;)$$

Figure 8.11. A double point resolution

type m for some m. Intuition of knot theorists has been that this extension by repeated differences corresponds to repeated derivatives, and hence finite type invariants can be thought of as 'polynomials'. The finite type invariants have a rich and interesting structure with strong connection to the Lie algebras. It remains a seminal open problem whether the finite type invariants separate knots, or at least recognize the unknot. All the knot invariants we mentioned in this book have expansions whose coefficients are finite type invariants constructed from Lie algebras.

A powerful tool in the study of an invariant V of type m is its *weight system* $V^{(m)}$, which gives its values on the diagrams with exactly m double points. The intuition is that these determine the 'constants of the leading terms of the polynomial'. Indeed, the weight system of the invariant V of type m determines V up to invariants of lower type. By definition, any change of the signs of the honest crossings in an m-singular diagram cannot change the value of an invariant of type m applied to it. Hence the weight system 'sees' only the *chord diagram* with m chords (an m-chord diagram) defined by the m-singular knot diagram (see Figure 8.12). The weight systems are directly characterized. Let

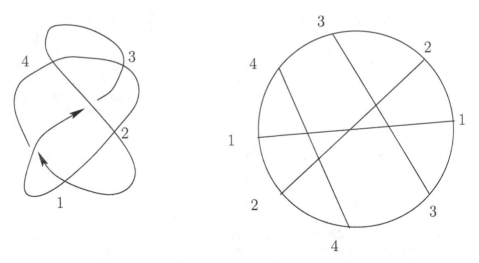

Figure 8.12. Figure 8 knot with double points and its chord diagram

Δ_m denote the space of all formal linear combinations with rational coefficients of $m-$chord diagrams. Let $\mathcal{A}_m^r$ be the quotient of Δ_m by all $4T$ relations (see Figure 8.13) and the FI relations $D = 0$, where D is a chord diagram with an isolated chord.

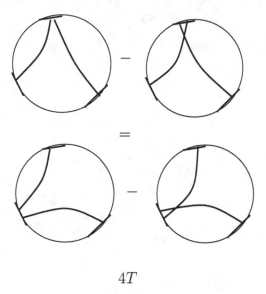

$$4T$$

Figure 8.13. 4T relation

Theorem 8.10.1. *If V is a rational valued type m invariant then its weight system $V^{(m)}$ defines a linear function on $\mathcal{A}_m^r$. Moreover, for any linear function W on $\mathcal{A}_m^r$ there is a rational valued type m invariant V so that $V^{(m)} = W$.*

This is the fundamental theorem of the theory of finite type invariants, see [B1]. Weight systems form a rich combinatorial structure whose connections to classical discrete mathematics have only started to be discovered. We have seen already an example of a weight system in the definition 7.5.8 of the U-polynomial. An interesting construction of Bar-Natan (see [B2]) relates the weight systems with the Four Color Theorem 2.10.3.

Chapter 9

2D Ising and dimer models

Since the solution of the 2-dimensional (planar) Ising problem was achieved by
Onsager, the physicists have been trying to reproduce his solution by more un-
derstandable methods. In the fifties and in the beginning of sixties two discrete
methods appeared: the *Pfaffian method* of Kasteleyn and independently Fisher,
Temperley, and the *path method* of Kac, Ward, Potts, Feynman and Sherman.
Both methods start by reducing the Ising partition function $Z(G, \beta)$ to the
generating function $\mathcal{E}(G, x)$ of even subsets of edges. This is accomplished by
van der Waerden's theorem (Theorem 6.3.1). The Pfaffian method seems to
be better known to discrete mathematicians. It further reduces $\mathcal{E}(G, x)$ to the
generating function $\mathcal{P}(G', x)$ of the perfect matchings (dimer arrangements) of
a graph G' obtained from G by a local operation at each vertex (see Section
6.2). It is important that these operations are locally planar, i.e., G' may be
embedded on the same surface as G.

9.1   Pfaffians, dimers, permanents

Let $G = (V, E)$ be a graph and let M, N be two perfect matchings of G. We
recall that $M \subset E$ is a matching if $e \cap e' = \emptyset$ for each pair e, e' of edges of M, and
a matching is perfect if its elements contain all the vertices of graph G. A cycle is
alternating with respect to a perfect matching M if it contains alternately edges
of M and out of M; each alternating cycle thus has an even length. We further
recall that Δ denotes the symmetric difference, $X \Delta Y = (X \setminus Y) \cup (Y \setminus X)$.
If m and N are two perfect matchings then $M \Delta N$ consists of vertex disjoint
alternating cycles.
Let C be a cycle of G of an even length and let D be an orientation of G. C is
said to be *clockwise even* in D if it has an even number of edges directed in D in
agreement with a chosen direction of traversal. Otherwise C is called *clockwise
odd*.

Definition 9.1.1. Let G be a graph with a weight function w on the edges.
Let D be an orientation of G. Let M be a perfect matching of G. For each

perfect matching P of G let $\text{sign}(D, M\Delta P) = (-1)^z$ where z is the number of clockwise even alternating cycles of $M\Delta P$. Moreover let

$$\mathcal{P}(D, M) = \sum_{P \text{ perfect matching}} \text{sgn}(D, M\Delta P)x^{w(P)}.$$

Let $G = (V, E)$ be a graph with $2n$ vertices and D an orientation of G. Denote by $A(D)$ the skew-symmetric matrix with the rows and the columns indexed by V, where $a_{uv} = x^{w(u,v)}$ in case (u, v) is an arc of D, $a_{u,v} = -x^{w(u,v)}$ in case (v, u) is an arc of D, and $a_{u,v} = 0$ otherwise.

Definition 9.1.2. The *Pfaffian* of $A(D)$ is defined as

$$\text{Pf}(A(D)) = \sum_P s^*(P)a_{i_1j_1} \cdots a_{i_nj_n},$$

where $P = \{\{i_1j_1\}, \cdots, \{i_nj_n\}\}$ is a partition of the set $\{1, \cdots, 2n\}$ into pairs, $i_k < j_k$ for $k = 1, \cdots, n$, and $s^*(P)$ equals the sign of the permutation $i_1j_1 \cdots i_nj_n$ of $12 \cdots (2n)$. Hence, each nonzero term of the expansion of the Pfaffian equals $x^{w(P)}$ or $-x^{w(P)}$ where P is a perfect matching of G. If $s(D, P)$ denotes the sign of the term $x^{w(P)}$ in the expansion, we may write

$$\text{Pf}(A(D)) = \sum_P s(D, P)x^{w(P)}.$$

The following theorem was proved by Kasteleyn.

Theorem 9.1.3. *Let G be a graph and D an orientation of G. Let P, M be two perfect matchings of G. Then*

$$s(D, P) = s(D, M)\text{sign}(D, M\Delta P).$$

Corollary 9.1.4.
$$\text{Pf}(A(D)) = s(D, M)\mathcal{P}(D, M).$$

The relevance of the Pfaffians in our context lies in the fact that the Pfaffian is a determinant-type function. The determinants are invariant under elementary row/column operations and these can be used in the Gaussian elimination to calculate a determinant. The Pfaffian may be computed efficiently by a variant of Gaussian elimination. Let A be an antisymmetric $2n \times 2n$ matrix. A *cross* of the matrix A is the union of a row and a column of the same index: the k-th cross is the following set of elements:

$$A_k = \{a_{ik}; 1 \leq i \leq 2n\} \cup \{a_{kj}; 1 \leq j \leq 2n\}.$$

Multiplying a cross A_k by a scalar α means multliplying each element of A_k by α.

Swapping crosses A_k and A_l means exchanging both the respective rows and columns. Another way of regarding the swap operation is that it exchanges the values of k and l in both of the index positions. The resulting matrix B is antisymmetric again;

Adding cross A_k to cross A_l means adding first the k-th row to the l-th one, and then adding the respective columns. The matrix remains antisymmetric.

These operations may be used to transform matrix A by at most $O(n^2)$ cross operations into a form where the Pfaffian can be determined trivially. Moreover, for graphs with some restrictive properties, e.g. for graphs with bounded genus, there are more efficient ways to perform the elimination. Apart of the Gaussian elimination, we also have the following classical theorem of Cayley.

Theorem 9.1.5.
$$(\mathrm{Pf}(A(D)))^2 = det(A(D)).$$

Kasteleyn introduced the following seminal notion:

Definition 9.1.6. A graph G is called Pfaffian if it has a *Pfaffian orientation*, i.e., an orientation such that all alternating cycles with respect to an arbitrary fixed perfect matching M of G are clockwise odd.

If G has a Pfaffian orientation D, then by Theorem 9.1.3 the signs $s(D, P)$ are all equal and $\mathcal{P}(G, x)$ is equal to $\mathrm{Pf}(A(D))$ up to a sign. Kasteleyn proved that each planar graph has a Pfaffian orientation.

Theorem 9.1.7. *Every topological planar graph has a Pfaffian orientation in which all inner faces are clockwise odd.*

Proof. Let G be a topological planar graph, and let M be a perfect matching in it. Without loss of generality we assume that G is 2-connected. Then, by Proposition 2.10.10, each face is bounded by a cycle. Starting with an arbitrary inner face, we can gradually construct an orientation D such that in D, each inner face is clockwise odd. Next we observe, e.g. by induction on the number of faces, that this orientation D satisfies: A cycle is clockwise odd if and only if it encircles an even number of vertices. However, each alternating cycle with a perfect matching in the complement must encircle an even number of vertices, and hence it is clockwise odd. $\qquad\square$

As a consequence we obtain the following theorem of Kasteleyn.

Theorem 9.1.8. *Each planar graph has an orientation D so that*

$$\mathcal{P}(G, x) = \mathrm{Pf}(A(D)).$$

Kasteleyn stated that for a graph of genus g, $\mathcal{P}(G, x)$ is a linear combination of 4^g Pfaffians. This was proved by Galluccio, Loebl and independently by Tesler. There were earlier partial results towards the proof by Regge and Zecchina. Tesler extended the result to the non-orientable surfaces. Galluccio and Loebl in fact proved the following compact formula.

Theorem 9.1.9. *If G is a graph of genus g then it has 4^g orientations $D_1, \cdots, D_{4^g}$ so that*

$$\mathcal{P}(G, x) = 2^{-g} \sum_{i=1}^{4^g} \text{sign}(D_i) Pf(A(D_i), x),$$

for well-defined $\text{sign}(D_i) \in \{1, -1\}$.

The proof can be found in [GL1]. Such a linear combination repair of a non-zero genus complication is a basic technique used both by mathematicians and physicists. The earliest work I have seen it in is by Kac and Ward; we will get to it in the next section. The next section also contains a theorem analogous to Theorem 9.1.9; there we will include the proof. Theorem 9.1.9 has attractive algorithmic consequences.

Corollary 9.1.10. *The Ising partition function $Z(G, \beta)$ can be determined efficiently for the topological graphs on an arbitrary surface of bounded genus. Also, the whole density function of the weighted edge-cuts, or weighted perfect matchings, may be computed efficiently for such graphs. Another well-known problem which is efficiently solvable for these graphs by the method of Theorem 9.1.9 is the exact matching problem: Given a positive integer k, a graph G and let the edges of G be colored by blue and red. It should be decided if there is a perfect matching with exactly k red edges.*
The efficiency is in the following sence: if we have integer weights, then the complexity is polynomial in the sum of the absolute values of the edge-weights.

We remark that a stronger notion of efficiency, where the complexity needs to be polynomial in the size of the graph plus the maximum of the logarithms of the edge-weights, is more customary. The existence of a polynomial algorithm in this sence is still open.
Curiously, there is no other polynomial method known to solve the max-cut problem alone even for the graphs on the torus. The method of Theorem 9.1.9 led to a useful implementation by Vondrák ([GLV1], [GLV2]).

Question 9.1.11. Is there an efficient combinatorial algorithm for the toroidal max-cut problem?

A lot of attention was given to the problem of characterizing graphs which admit a Pfaffian orientation. The problem of recognizing the Pfaffian bipartite graphs goes implicitly back to 1913, when Pólya asked for a characterization of convertible matrices (this is the 'Pólya scheme'). A matrix A is *convertible* if one can change some signs of its entries to obtain a matrix B such that $Per(A) = \det(B)$. A polynomial-time algorithm to recognize the Pfaffian bipartite graphs (this problem is equivalent to the Pólya problem described above) has been obtained by McCuaig, Robertson, Seymour and Thomas. For the recognition of the Pfaffian graphs embeddable on an arbitrary 2-dimensional surface, there is a polynomial algorithm by Galluccio and Loebl (using Theorem 9.1.9). Theorem 9.1.9 can also be used in a straightforward way to complete the Pólya scheme.

Corollary 9.1.12. *For each matrix A there are matrices B_i, $i = 1, \cdots, 4^g$, obtained from A by changing signs of some entries, so that $\mathrm{Per}(A)$ is an alternating sum of the $\det(B_i)$'s. The parameter g is the genus of the bipartite graph determined by the non-zero entries of A.*

Several researchers (Hammersley, Heilmann, Lieb, Godsil, Gutman) noticed that $\mathrm{Per}(A)$, A a general complex matrix, is equal to the expectation of $(\det(B))^2$, where B is obtained from A by taking the square root of the minimal argument of each non-zero entry and then multiplying each non-zero entry by an element of $\{1, -1\}$ chosen independently uniformly at random. This leads to a Monte-Carlo algorithm for estimating the permanent (see Karmarkar, Karp, Lipton, Lovász and Luby [KKLLL] for the rate of convergence analysis).

Theorem 9.1.13. *Let A be a matrix and let B be the random matrix obtained from A by taking the square root of minimal argument of each non-zero entry and then multiplying each non-zero entry by an element of $\{1, -1\}$ chosen independently uniformly at random. Then $\mathbb{E}((\det(B))^2) = \mathrm{Per}(A)$.*

Proof. Since $\det(B) = \sum_\pi \mathrm{sign}(\pi) \prod_i B_{i\pi(i)}$, we have

$$(\det(B))^2 = \sum_{(\pi_1, \pi_2)} \mathrm{sign}(\pi_1)\mathrm{sign}(\pi_2)) \prod_i B_{i\pi_1(i)} B_{i\pi_2(i)} =$$

$$\sum_\pi \mathrm{sign}(\pi)^2 \prod_i B_{i\pi(i)}^2 +$$

$$\sum_{(\pi_1, \pi_2); \pi_1 \neq \pi_2} \mathrm{sign}(\pi_1)\mathrm{sign}(\pi_2)) \prod_i B_{i\pi_1(i)} B_{i\pi_2(i)} =$$

$$\mathrm{Per}(A) + \sum_{(\pi_1, \pi_2); \pi_1 \neq \pi_2} \mathrm{sign}(\pi_1)\mathrm{sign}(\pi_2)) \prod_i B_{i\pi_1(i)} B_{i\pi_2(i)}.$$

It remains to show that the expectation of the last sum is zero. Let A be an $n \times n$ matrix and let $\pi_1 \neq \pi_2$ be two permutations of n. We can associate with them a graph $G(\pi_1, \pi_2)$. Its vertex-set is the set of all pairs (i, j) for $j = \pi_1(i)$ or $j = \pi_2(i)$. Two vertices $(i, j), (i', j')$ are connected by an edge if and only if $i = i'$ or $j = j'$. We recall that $c(G)$ denotes the number of the connected components of G.

Clearly, each $G(\pi_1, \pi_2)$ has at least one edge, and the non-empty components of each $G(\pi_1, \pi_2)$ are cycles of an even length. Let $\mathcal{G}$ be the set of all such graphs $G(\pi_1, \pi_2)$ for some $\pi_1 \neq \pi_2$. If $G \in \mathcal{G}$ then we let $eq(G) = \{(\pi_1, \pi_2) : G = G(\pi_1, \pi_2)\}$. We observe that $|eq(G)| = 2^{c(G)}$. Finally let us denote by $(ij)(G)$ an arbitrary vertex of G which belongs to a cycle. Now, we can write

$$\sum_{(\pi_1, \pi_2); \pi_1 \neq \pi_2} \mathrm{sign}(\pi_1)\mathrm{sign}(\pi_2)) \prod_i B_{i\pi_1(i)} B_{i\pi_2(i)} =$$

$$\sum_{G \in \mathcal{G}} \sum_{(\pi_1, \pi_2) \in eq(G)} \mathrm{sign}(\pi_1)\mathrm{sign}(\pi_2)) \prod_i B_{i\pi_1(i)} B_{i\pi_2(i)} =$$

$$\sum_{G \in \mathcal{G}} B_{(ij)(G)} y(G),$$

where $y(G)$ is a random variable independent of $B_{(ij)(G)}$. Since the expectation of $B_{(ij)(G)}$ is equal to zero, the proof is finished.

$\square$

However, for the matrices with $0, 1$ entries, there is something better. Jerrum, Sinclair and Vigoda constructed a *fully polynomial randomized approximation scheme* (FPRAS, in short) for approximating permanents of matrices with *nonnegative entries*. Briefly, a FPRAS for the permanent is an algorithm which, when given as input an $n \times n$ nonnegative matrix A together with an accuracy parameter $\epsilon \in (0, 1]$, outputs a number Z (a random variable of the coins tossed by the algorithm) such that

$$\text{Prob}[(1 - \epsilon)Z \leq \text{Per}(A) \leq (1 + \epsilon)Z] \geq \frac{3}{4}$$

and runs in time polynomial in $n, \sum |log(A_{ij})|$ and ϵ^{-1}. The probability $3/4$ can be increased to $1 - \delta$ for any desired $\delta \in (0, 1]$ by outputting the median of $O(\log \delta^{-1})$ independent trials.

9.2   Products over aperiodic closed walks

The following solution to the 2-dimensional Ising model has been developed by Kac, Ward and Feynman. This theory is closely related to that of Section 7.1. Let $G = (V, E)$ be a planar topological graph. It is convenient to associate a variable x_e instead of a weight to each edge e. If $e \in E$ then a_e will denote the orientation of e and a_e^{-1} will be the reversed orientation. We let $x_a = x_e$ for each orientation a of e. A circular sequence $p = v_1, a_1, v_2, a_2, ..., a_n, (v_{n+1} = v_1)$ is called a *prime reduced cycle*, if the following conditions are satisfied: $a_i \in \{a_e, a_e^{-1} : e \in E\}$, $a_i \neq a_{i+1}^{-1}$ and $(a_1, ..., a_n) \neq Z^m$ for some sequence Z and $m > 1$. We let $X(p) = \prod_{i=1}^{n} x_{a_i}$ and if each degree of G is at most 4 then we let $W(p) = (-1)^{\text{rot}(p)} X(p)$ where $\text{rot}(p)$ was defined in Chapter 5.

If $E' \subset E$ then we also let $X(E') = \prod_{e \in E'} x_e$. There is a natural equivalence on the prime reduced cycles: p is equivalent to reversed p. Each equivalence class has two elements and will be denoted by $[p]$. We let $W([p]) = W(p)$ and note that this definition is correct since equivalent walks have the same sign. The following theorem was proposed by Feynman and proved by Sherman. It provides, for a planar graph G, an expression for the generating function $\mathcal{E}(G, x)$ of the even sets of edges (see Section 2.1), in terms of the Ihara-Selberg function of G (see Definition 7.1.1).

Theorem 9.2.1. *Let G be a planar topological graph with each degree even and at most 4. Then*

$$\mathcal{E}(G, x) = \prod (1 - W([p])),$$

where we denote by $\prod(1-W([p])$ the formal product of $(1-W([p])$ over all equivalence classes of prime reduced cycles of G (the formal product was considered in Chapter 7).

Note that the product is infinite even for a very simple graph consisting of one vertex and two loops.

When each transition between a pair of directed edges is decorated by its rotation contribution (see Section 5.3), Theorem 9.2.1 implies that $\mathcal{E}^2(G,x)$ becomes an Ihara-Selberg function (see Section 7.1). Hence we get the following corollary, whose statement (and incorrect proof) by Kac and Ward was in fact the starting point of the whole path approach.

Theorem 9.2.2. *Let G be a topological planar graph with all degrees even and at most 4. Then $\mathcal{E}^2(G,x)$ equals the determinant of the transition matrix between directed edges; each transition is decorated by its rotation contribution.*

Theorem 9.2.1 is formulated for those topological planar graphs where each degree is even and at most 4. It is not difficult to reduce $\mathcal{E}(G,x)$, G a general topological planar graph, to this case: First we make each degree even by doubling each edge. If we set the variables of the new edges to zero then each term containing a contribution of at least one new edge disappears. Next we make each non-zero degree equal to 2 or 4 as follows. We replace each vertex v with incident edges $e_1, ..., e_{2k}$, $k > 2$, listed in the circular order given by the embedding of G in the plane, by a path P of $2k - 2$ vertices. We set the variables of the edges of P equal to 1. Next we double each edge of the unique perfect matching of P and set the variables of the new edges to zero. Finally we join the edges $e_1, ..., e_{2k}$ to the vertices of the auxiliary path so that the order is preserved along the path and each degree is four: there is a unique way to do that.

In order to prove Theorem 9.2.1, Sherman formulated and proved the following generalization which we now state. Let v be a vertex of degree 4 of G and let p be an aperiodic closed walk of G. We say that p satisfies the *crossover condition* at v if the way p passes through v is consistent with the crossover pairing of the four edges incident with v.

Let U be a subset of vertices of degree 4. An even subset $E' \subset E$ is called *acceptable* for U if, for each $u \in U$ and for both pairs of edges incident with u and paired by the crossover pairing at u, if E' contains one edge of the pair then it also contains the other one.

Theorem 9.2.3. *Let $G = (V, E)$ be a topological planar graph where each degree is even and at most 4. Let U be a subset of vertices of G of degree 4. Let $\prod'_{G,U}(1-W([p]))$ denote the product over all equivalence classes of the aperiodic closed walks of G which satisfy the crossover condition at each $u \in U$. Then*

$$\prod\nolimits'_{G,U}(1 - W([p])) = \sum(-1)^{c(E')}X(E'),$$

where the sum is over all acceptable even subsets $E' \subset E$ and $c(E')$ is equal to the number of vertices of U such that E' contains all four edges incident with it.

The proof proceeds in two steps. First we show that, when the infinite product is expanded as a sum of monomials of variables, the coefficient corresponding to $X(E')$, for any E' acceptable for U, is equal to $(-1)^{c(E')}$. In the second step we show that all the remaining coefficients are zero.

Proposition 9.2.4. *Let E' be acceptable for U. If $\prod'_{G,U}(1-W([p]))$ is expanded as a sum of monomials of variables then the coefficient of $X(E')$ is equal to $(-1)^{c(E')}$.*

Proof. By induction on the number of vertices of non-zero degree in E'. If E' has just one vertex then it consists of one loop e or two loops e, f and $c(E')$ equals zero or one. If E' consists of one loop only then $\prod'_{G,U}(1 - W([p])) = (1+x_e)\times$ product of terms which cannot influence the coefficient at $X(E')$. If E' consists of two loops and $c(E') = 0$ then $\prod'_{G,U}(1 - W([p]))$ equals $(1 + x_e)(1 + x_f)(1 + x_e x_f)(1 - x_e x_f)\times$ product of terms which cannot influence the coefficient at $X(E')$. Finally let $c(E') = 1$ and E' consist of two loops. $\prod'_{G,U}(1 - W([p]))$ equals $(1 - x_e x_f)\times$ product of terms which cannot influence the coefficient at $X(E')$. Hence the base of the induction is verified.

Now we assume the statement is true for all acceptable subsets of edges with $n \geq 1$ vertices of non-zero degree. Let E' be an acceptable subgraph with $n+1$ vertices of non-zero degree. A vertex v will be called *free* if it does not contribute to $c(E')$, i.e., if v has degree 2 in E' or $v \notin U$. Let $k = n + 1 - c(E')$ be the number of free vertices.

We continue by induction on k. First let $k = 0$, i.e., each vertex of non-zero degree in E' has degree 4 and belongs to U. The crossover conditions cause that there is a unique decomposition of E' into prime reduced cycles $p_1, ..., p_r$ such that $X(E') = \prod_{i=1}^{r} X(p_i)$. If $r = 1$ then by Observation 5.3.4, $(-1)^{\mathrm{rot}(p_1)} = (-1)^{c(E')}$. If $r > 1$ then $\prod_{i=1}^{r}(-1)^{\mathrm{rot}(p_i)} = (-1)^{c(E')}$ since any two of the p_i's mutually intersect in an even number of vertices, and each vertex contributes to $c(E')$.

Hence let $k > 0$ and the statement holds for all acceptable subsets with less than k free vertices. If all free vertices have degree 2 in E' then we may proceed as in the case $k = 0$. Hence let v be a free vertex of E' of degree four in E'. We denote the edges incident with v by north, east, south and west according to the cyclic order induced by the embedding in the plane. We partition the prime reduced cycles of G which satisfy the crossover conditions at the vertices of U into four classes. Classes I,II,III contain prime reduced cycles that have an edge incident with v, and:

class I contains the prime reduced cycles that are consistent with west-north and east-south pairing,

class II contains the prime reduced cycles that are consistent with west-south and east-north pairing,

class III contains the prime reduced cycles that are consistent with north-south and east-west pairing , and finally

class IV contains the prime reduced cycles that do not contain any edge incident with v. Suppose $p \in I$ and $q \in II$. Then the product $W[p]W[q]$ contains

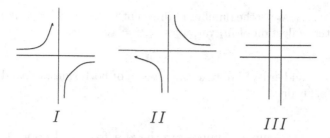

Figure 9.1. Classes I- III

a variable with the exponent bigger than 1. Hence it can make no contribution to $X(E')$. The same is true for II, III and I, II. Hence, if $\prod'_{G,U}(1 - W([p])$ is expanded as a sum, the coefficient of $X(E')$ is the sum of the corresponding coefficients in $I \times IV$, $II \times IV$ and $III \times IV$.

The contribution to $I \times IV$ can be regarded as the coefficient of $X(E'')$ in $\prod'_{G',U}(1 - W([p]))$ where G' and E'' are obtained from G and E' by deleting vertex v and by identifying the west, north edges into one edge, and the east, south edges into one edge. Analogously, we can treat the case $II \times IV$. Hence by the induction assumption the sum of the contributions from $I \times IV$ and $II \times IV$ is $2(-1)^{c(E')}$. The contribution to $III \times IV$ can be regarded as coming from $\prod'_{G,U\cup\{v\}}(1 - W([p]))$, i.e. one additional cross-over condition is imposed, on vertex v. Using the induction assumption again (this time for k) we get that this contribution is equal to $(-1)^{c(E')+1}$.

Summarizing when the product $\prod'(1 - W([p]))$ is expanded as a sum, the coefficient of $X(E')$ is equal to $2(-1)^{c(E')} + (-1)^{c(E')+1}$, which we wanted to show.

□

To finish the proof of Theorem 9.2.3, we need to show that the remaining coefficients of the expansion of the infinite product are all equal to zero. We observe that the remaining coefficients belong to terms which are products of variables where at least one of the exponents is greater than 1.

We temporarily consider $\prod'_{G,U}(1 - W(p))$, where now the product is over prime reduced cycles and so it is the square of the original $\prod'_{G,U}(1 - W[p]))$. Let $a_1 > a_1^{-1} > ... > ...$ be a linear order of orientations of the edges of G.

Let A_1 be the set of all prime reduced cycles p such that a_1 appears in p. Each $p \in A_1$ has a unique factorization into words $(W_1, ..., W_k)$ each of which starts with a_1 and has no other appearance of a_1. Some of these words contain a_1^{-1} and some do not. We will need a lemma on coin arrangements stated below. The lemma was proved by Sherman. We present a proof based on the Witt identity from combinatorial group theory.

Witt Identity: Let $z_1, \ldots, z_k$ be commuting variables. Then

$$\prod_{m_1,...,m_k \geq 0} (1 - z_1^{m_1} \cdots z_k^{m_k})^{M(m_1,...,m_k)} = 1 - z_1 - z_2 - ... - z_k,$$

where $M(m_1, \ldots, m_k)$ is the number of different non-periodic circular sequences made from the collection of m_i variables z_i, $i = 1, \cdots, k$.

Proof. (of Witt's identity) We take the inverse of both sides, expand and apply the Lyndon's Theorem 7.1.5.

□

Here comes the lemma. Suppose we have a fixed collection of N objects of which m_i are of ith kind, $i = 1, \ldots, n$. Let b_k be the number of exhaustive unordered arrangements of these symbols into k disjoint, nonempty, circularly ordered sets such that no two circular orders are the same and none are periodic. For example suppose we have 10 coins of which 3 are pennies, 4 are nickles and 3 are quarters. The arrangement $\{(p, n), (n, p), (p, n, n, q, q, q)\}$ is not counted in b_3 since (p, n) and (n, p) represent the same circular order.

Lemma 9.2.5. *(On coin arrangements) If $N > 1$ then $\sum_{i=1}^{N}(-1)^i b_i = 0$.*

Proof. The lemma follows immediately if we expand the LHS of the Witt identity and collect the terms where the sums of the exponents of z_i's are the same.

□

Proposition 9.2.6. $\prod_{p \in A_1}(1 - W(p)) = 1 + x_{a_1}d_{11}$ *where d_{11} is a formal (possibly infinite) sum of monomials none of which has x_{a_1} as a factor.*

Proof. First we note that the additivity of rotation implies the following fact: if p_1, p_2 are two prime reduced cycles both containing a_1 and p_1p_2 is also prime reduced then $(-1)^{\text{rot}(p_1p_2)} = (-1)^{\text{rot}(p_1)+\text{rot}(p_2)}$.
Let D be a monomial summand in the expansion of $\prod_{p \in A_1}(1 - W(p))$. Hence D is a product of finitely many $W(p), p \in A_1$. Each $p \in A_1$ has a unique factorization into words $(W_1, ..., W_k)$ each of which starts with a_1 and has no other appearance of a_1. Each word may appear several times in the factorization of p, and also in the factorization of different prime reduced cycles of A_1. Let $B(D)$ be the set-system of all the words (with repetition) appearing in the factorizations of the prime reduced cycles of D. It follows from the lemma on coin arrangements that the sum of all monomial summands D in the expansion of $\prod_{p \in A_1}(1 - W(p))$, which have the same $B(D)$ of more than one element, is zero. Hence the monomial summands D which survive in the expansion of $\prod_{p \in A_1}(1 - W(p))$ all have $B(D)$ consisting of exactly one word. This word may but need not contain a_1^{-1}. However, only the summands with their word NOT containing a_1^{-1} survive, by the following observation: If $b, c_1, ..., c_k$ are walks that contain neither a_1 nor a_1^{-1} then

$$W(a_1ba_1^{-1}c_1a_1^{-1}c_2...a_1^{-1}c_k) + W(a_1b^{-1}a_1^{-1}c_1a_1^{-1}c_2...a_1^{-1}c_k)+$$
$$W(a_1ba_1^{-1}c_1a_1^{-1}c_2...a_1^{-1}c_k^{-1}) + W(a_1b^{-1}a_1^{-1}c_1a_1^{-1}c_2...a_1^{-1}c_k^{-1}) = 0.$$

□

Analogously, let A_2 be the set of all prime reduced cycles p such that a_1^{-1} appears in p. Possibly $A_1 \cap A_2 \neq \emptyset$. Analogously as for $p \in A_1$, each $p \in A_2$ has a unique factorization into words $(W_1, ..., W_k)$ each of which starts with a_1^{-1} and has no other appearance of a_1^{-1}. Some of these words contain a_1 and some do not. The following proposition may be proved in exactly the same way as Proposition 9.2.6.

Proposition 9.2.7. *Let* A_1, A_2 *be as above. Then*

$$\prod_{p \in A_2} (1 - W(p)) = \prod_{p \in A_1 \backslash A_2} (1 - W(p)) = \prod_{p \in A_2 \backslash A_1} (1 - W(p)) = \prod_{p \in A_1} (1 - W(p)).$$

Let B be the set of prime reduced cycles in which neither a_1 nor a_1^{-1} appears. We may write

$$\prod_{p \in B} (1 - W(p)) = (1 + d_{12})^2,$$

where d_{12} is a formal sum of monomials, none of which has x_{a_1} as a factor. In $\prod_{p \in A_1} (1 - W(p)) \times \prod_{p \in A_2} (1 - W(p)) = (1 + x_{a_1} d_{11})^2$, the prime reduced cycles from $A_1 \cap A_2$ have been counted doubly, while the prime reduced cycles from $A_1 \backslash A_2$ and $A_2 \backslash A_1$ have been counted only once. Hence

$$\left(\prod_{p \in (A_1 \cup A_2)} (1 - W(p)) \right)^2 =$$

$$\prod_{p \in A_1} (1 - W(p)) \times \prod_{p \in A_2} (1 - W(p)) \times \prod_{p \in A_1 - A_2} (1 - W(p)) \times \prod_{p \in A_2 - A_1} (1 - W(p)) =$$

$$(1 + x_{a_1} d_{11})^4.$$

Proof. (of Theorem 9.2.3)

$$\left(\prod{}'_{G,U} (1 - W([p])) \right)^2 = \prod{}'_{G,U} (1 - W(p)) =$$

$$\prod_{p \in (A_1 \cup A_2)} (1 - W(p)) \times \prod_{p \in B} (1 - W(p)) =$$

$$(1 + x_{a_1} d_{11})^2 (1 + d_{12})^2,$$

and

$$\prod{}'_{G,U} (1 - W([p])) = (1 + x_{a_1} d_{11})(1 + d_{12}).$$

Thus, there are no monomial summands having factors $x_{a_1}^n$, $n \geq 2$. The same argument disposes of the summands with factors $x_{a_i}^n$, $i \neq 1, n \geq 2$.

□

Theorem 9.2.3 can be used to express $\mathcal{E}(G, x)$ for general graphs as a linear combination of infinite products. A useful trick to obtain explicit formulas is to base such a linear combination on the genus. This we explain next. Let us first consider the graphs embeddable on torus (they are usually called *toroidal graphs*). We will again assume that each degree is even and at most 4. Let us take a natural representation of the torus as a rectangle with opposite edges identified. The edges of the original rectangle form two cycles on the torus. Let us call them the *vertical cycle*, and the *horizontal cycle*. Let G be a topological toroidal graph such that no vertex belongs to the horizontal or to the vertical cycle. If p is a prime reduced cycle of G, then let $h(p)$ denote the number of times p crosses the horizontal cycle, and let $v(p)$ denote the number of times p crosses the vertical cycle. The notation $h(E')$ and $v(E')$ is also used for even subsets E' of G. How do we define $\mathrm{rot}(p)$ on the torus? We unglue the edges of the rectangle which represents the torus. Hence each rectangle edge crossing now corresponds to 'leaving' the rectangle and 'coming back' to the rectangle by the opposite rectangle edge. If we draw all this in the plane, we get $h(G)v(G)$ crossings of the curves representing the edges of G. Let G' be the graph obtained from G by introducing a vertex to each such intersection. Note that G' is properly drawn in the plane and each degree of G' is even and at most four. Let us call the new vertices *special* and note that each special vertex has degree four in G'. Further note that each prime reduced cycle p of G corresponds to the prime reduced cycle p' of G' which satisfies the crossover condition at each special vertex. We let

$$(-1)^{\mathrm{rot}(p)} = (-1)^{h(p)+v(p)}(-1)^{\mathrm{rot}(p')}.$$

Finally we let

$$W_h(p) = (-1)^{h(p)}W(p),$$

$$W_v(p) = (-1)^{v(p)}W(p)$$

and

$$W_{h,v}(p) = (-1)^{h(p)+v(p)}W(p).$$

Hence

$$W([p']) = W_{h,v}([p]).$$

Theorem 9.2.9 and in particular Theorem 9.2.12 are based on the following curious lemma.

Lemma 9.2.8. *Let R be the set of all $0,1$-vectors of length $2n$ and let a be an arbitrary integer vector of length $2n$. Then*

$$2^{-n}(-1)^{\sum_{i=1}^{n} a_{2i-1}a_{2i}}\left(\sum_{r \in R}(-1)^{ra}(-1)^{s(r)}\right) = 1,$$

where $s(r)$ denotes the number of i such that $r_{2i-1} = r_{2i} = 1$.

Proof. We proceed by induction on n. The initial case $n = 1$ may be easily checked by hand. Next assume that Lemma 9.2.8 is true for n and we want to prove it for $n + 1$. Let R' be the set of all $0, 1$-vectors of length $2(n + 1)$ and let a' be an arbitrary integer vector of length $2(n + 1)$. Let a denote the initial part of a' of length $2n$. Then

$$2^{-n-1}(-1)^{\sum_{i=1}^{n+1} a'_{2i-1}a'_{2i}} \left(\sum_{r \in R'} (-1)^{ra'}(-1)^{s(r)} \right) =$$

$$2^{-1}(-1)^{a'_{2n+1}a'_{2n+2}} \alpha[(-1)^{a'_{2n+1}} + (-1)^{a'_{2n+2}} - (-1)^{a'_{2n+1}+a'_{2n+2}} + 1],$$

where

$$\alpha = 2^{-n}(-1)^{\sum_{i=1}^{n} a_{2i-1}a_{2i}} \left(\sum_{r \in R} (-1)^{ra}(-1)^{s(r)} \right).$$

By induction assumption we have that $\alpha = 1$ and applying again the first step of the induction, we find that the lemma holds. $\square$

Theorem 9.2.9. *If $G = (V, E)$ is a toroidal graph where each degree is even and at most four, then*

$$\mathcal{E}(G, x) =$$

$$1/2 \left(\prod(1 - W_h([p])) + \prod(1 - W_v([p])) + \prod(1 - W_{h,v}([p])) - \prod(1 - W([p])) \right),$$

where $\prod$ is the product over all equivalence classes of prime reduced cycles of G.

Proof. Using Theorem 9.2.3 we get that

$$\prod(1 - W_{h,v}([p])) = \prod{}'(1 - W([p'])) = \sum(-1)^{h(E')v(E')}X(E'),$$

where the sum goes over all acceptable subgraphs E' of G', i.e. over all even subgraphs of G. Hence also

$$\prod(1 - W_v([p])) = \sum(-1)^{h(E')v(E')+h(E')}X(E'),$$

$$\prod(1 - W_h([p])) = \sum(-1)^{h(E')v(E')+v(E')}X(E'),$$

and

$$\prod(1 + W([p])) = \sum(-1)^{h(E')v(E')+h(E')+v(E')}X(E').$$

Let E' be an arbitrary even subset of G. Then the coefficient of $X(E')$ in

$$1/2 \left(\prod(1 - W_h([p])) + \prod(1 - W_v([p])) + \prod(1 - W_{h,v}([p])) - \prod(1 - W([p])) \right)$$

equals

$$1/2(-1)^{h(E')v(E')} \left((-1)^{h(E')} + (-1)^{v(E')} - (-1)^{h(E')+v(E')} + 1 \right) = 1,$$

by Lemma 9.2.8. $\square$

Using the machinery of *g-graphs* (see Definition 9.2.11), we can write down a formula for general graphs. The machinery is based on the following representation of orientable surfaces.

Definition 9.2.10. A highway surface S_g consists of a *base* B_0 and $2g$ *bridges* B_j^i, $i = 1, ..., g$ and $j = 1, 2$, where

(i) B_0 is a convex $4g$-gon with vertices $a_1, ..., a_{4g}$ numbered clockwise;

(ii) B_1^i, $i = 1, \cdots, g$, is a 4-gon with vertices $x_1^i, x_2^i, x_3^i, x_4^i$ numbered clockwise. It is glued with B_0 so that the edge $[x_1^i, x_2^i]$ of B_1^i is identified with the edge $[a_{4(i-1)+1}, a_{4(i-1)+2}]$ of B_0 and the edge $[x_3^i, x_4^i]$ of B_1^i is identified with the edge $[a_{4(i-1)+3}, a_{4(i-1)+4}]$ of B_0;

(iii) B_2^i, $i = 1, \cdots, g$, is a 4-gon with vertices $y_1^i, y_2^i, y_3^i, y_4^i$ numbered clockwise. It is glued with B_0 so that the edge $[y_1^i, y_2^i]$ of B_2^i is identified with the edge $[a_{4(i-1)+2}, a_{4(i-1)+3}]$ of B_0 and the edge $[y_3^i, y_4^i]$ of B_2^i is identified with the edge $[a_{4(i-1)+4}, a_{4(i-1)+5(mod4g)}]$ of B_0.

We remark that in Definition 9.2.10 we denote by $[a, b]$ edges of polygons and not edges of graphs. The usual representation in the space of an orientable surface S of genus g may then be obtained from S_g by the following operation: for each bridge B, glue together the two segments which B shares with the boundary of B_0, and delete B.

Definition 9.2.11. A graph G is called a *g-graph* if it is embedded on S_g so that all the vertices belong to the base B_0, and each time an edge intersects a bridge, it crosses it completely.

This is analogous to the situation described earlier for the torus: we can imagine that we contract all the bridges (and get a usual representation of an orientable surface of genus g), draw our graph there, and then split the bridges back. The resulting drawing is a *g-graph* on S_g. If G is a *g-graph* and p is a prime reduced cycle of G then we denote by $a(p)$ the vector of length $2g$ such that $a(p)_{2(i-1)+j}$ equals the number of times p crosses bridge B_j^i, $i = 1, ..., g$, $j = 1, 2$. Similarly we will use the notation $a(E')$ where E' is an even subset of G.

Note that any graph G can be embedded as a *g-graph* where g is genus of G. As before, we only need to consider *g-graphs* that have all degrees even and at most four (by a remark after Theorem 9.2.1). We define $(-1)^{\text{rot}(p)}$ analogously as for the torus: We consider G embedded in the plane by the projection of the bridges B_j^i outside B_0. We get $\sum_{i=1}^g a(G)_{2i-1} a(G)_{2i}$ crossings of the curves representing the edges of G. Let G' be the graph obtained from G by introducing a vertex to each such intersection. Note that G' is a topological planar graph, and each degree of G' is even and at most four. Let us call the new vertices *special* and note that each special vertex has degree 4 in G'. Each non-periodic closed walk p of G corresponds to the prime reduced cycle p' of G' which satisfies the crossover condition at each special vertex. Let J denote the vector $(1, \ldots, 1)$ of all 1's. We define $(-1)^{\mathrm{rot}(p)}$ by

$$(-1)^{\mathrm{rot}(p')} = (-1)^{Ja(p)}(-1)^{\mathrm{rot}(p)}.$$

Let $R(g)$ denote the set of all $0,1$-vectors of length $2g$. For $r \in R(g)$ we let $W_r([p]) = (-1)^{ra(p)}W([p])$. Hence $W([p']) = W_J([p])$.

Theorem 9.2.12. *If $G = (V, E)$ is a g-graph where each degree is even and at most four, then*

$$\mathcal{E}(G, x) = 2^{-g} \sum_{r \in R(g)} (-1)^{s(J-r)} \prod (1 - W_r([p])),$$

where $\prod$ is the formal infinite product over all equivalence classes of prime reduced cycles of G.

Proof. We proceed as in the proof of Theorem 9.2.9. Using Theorem 9.2.3 we get

$$\prod (1 - W_J([p])) = \prod{}'(1 - W([p'])) = \sum (-1)^{\sum_{i=1}^{g} a(E')_{2i-1}a(E')_{2i}} X(E'),$$

where the sum is over all acceptable subsets E'' of G', i.e., over all even subsets of G. Hence for $r \in R(g)$ we have

$$\prod (1 - W_r([p])) = \sum (-1)^{\sum_{i=1}^{g} a(E')_{2i-1}a(E')_{2i} + (J-r)a(E')} X(E'),$$

where the sum is over all even subsets E' of G. Let E' be an arbitrary even subset of G. Then the coefficient of $X(E')$ in

$$2^{-g} \sum_{r \in R(g)} (-1)^{s(J-r)} \prod (1 - W_r([p]))$$

is equal to

$$2^{-g}(-1)^{\sum_{i=1}^{g} a(E')_{2i-1}a(E')_{2i}} \sum_{r \in R(g)} (-1)^{(J-r)a(E')}(-1)^{s(J-r)} = 1,$$

by Lemma 9.2.8, since we can replace r by $J - r$ in the summation. $\square$

Bibliography

[AKS]   M. Ajtai, J. Komlós, E. Szemerédi, *On a conjecture of Loebl*, In: Proc.7th International Conf. on Graph Theory, Combinatorics and Algorithms, Wiley, New York (1995) 1135–1146.

[A]   J.W. Alexander, *Topological invariants of knots and links*, Trans. Amer. Math. Soc. **30** (1928) 275–306.

[AH]   K. Appel, W. Haken, *Every planar map is four colorable*, Contemp. Math. **98** (1989).

[B1]   D. Bar-Natan, *On the Vassiliev knot invariants*, Topology, **34** (1995) 423–172.

[B2]   D. Bar-Natan, *Lie algebras and the four color theorem*, Combinatorica, **17** (1997) 43–52.

[B]   H. Bass, *The Ihara-Selberg zeta function of a tree lattice*, Intern. J. Math. **3** (1992) 717–797.

[BRJ]   R.J. Baxter, *Exactly solved models in statistical physics*, Academic Press London (1982).

[BB]   B. Bollobas, *Modern graph theory*, Springer-Verlag (1998).

[BR]   B. Bollobas and O. Riordan, *A polynomial of graphs on surfaces*, Math. Ann. **323** (2002) 81–96.

[BLP]   R.A. Brualdi, M. Loebl and O. Pangrac, *Perfect Matching Preservers*, Electronic J. of Combinatorics (2006).

[BT]   T.H. Brylawski, *Intersection theory for graphs*, J. Combin. Theory Ser.B **30** (1981) 233–246.

[CDL]   S.V. Chmutov, S.V. Duzhin, S.K. Lando, *Vassiliev knot invariants I-III*, Advances Soviet Math. **21** (1994).

[E1]   J. Edmonds, *Paths, trees, and flowers*, Canadian J. of Mathematics **17** (1965) 449–467.

[E2]   J. Edmonds, *Minimum partition of a matroid into independent subsets*, Journal of Research National Bureau of Standards Section B **69** (1965) 67–72.

[GJ]   J. de Gier, *Loops, matchings and alternating-sign matrices*, Discrete Mathematics **298** (2005) 365–388.

[EM]   N.M. Ercolani, K.D.T.-R. McLaughlin, *Asymptotics of the partition function for random matrices via Riemann-Hilbert techniques and application to graphical enumeration*, Internat. Math. Res. Notices **14** (2003) 755–820.

[FOM]   H. de Fraysseix, P. Ossona de Mendez, *On a characterisation of Gauß codes*, Discrete and Computational Geometry **22** (1999) 287- 295.

[F]   P. Di Francesco, *2D quantum gravity, matrix models and graphs combinatorics*, survey (2004).

[FZ]   D. Foata and D. Zeilberger, *A combinatorial proof of Bass's evaluation of the Ihara-Selberg zeta function for graphs*, Transactions Amer. Math. Soc. **351** (1999) 2257–2274.

[FK]   C.M. Fortuin and P.W. Kasteleyn, *On the random-cluster model I. Introduction and relation to other models*, Physica **57** (1972) 536-564.

[GLV1]   A. Galluccio, M. Loebl and J. Vondrák, *Optimization via enumeration: a new algorithm for the max cut problem*, Mathematical Programming **90** (2001) 273-290.

[GLV2]   A. Galluccio, M. Loebl and J. Vondrák, *A new algorithm for the ising problem: partition function for finite lattice graphs*, Physical Review Letters **84** (2000) 5924-5927.

[GL1]   A. Galluccio and M. Loebl, *A Theory of Pfaffian orientations I*, Electronic Journal of Combinatorics **6** (1999).

[GL2]   A. Galluccio and M. Loebl, *A Theory of Pfaffian orientations II* Electronic Journal of Combinatorics **6** (1999).

[GL]   S. Garoufalidis, M. Loebl, *A non-commutative formula for the colored Jones function*, Mathematische Annalen **336** (2006) 867- 900.

[GJ]   M.R. Garrey, D.S. Johnson, *Computers and intractability:A guide to the theory of NP-completeness*, Freeman, San Francisco (1979).

[G]   C.F. Gauß, *Werke VIII*, Teubner, Leipzig, (1900) 282- 286.

[GC]   C. Green, *Weight Enumeration and the Geometry of Linear Codes*, Studies in Applied Mathematics **55** (1976) 119–128.

[JKS]   F. Jaeger, L. Kauffman and P. Saleur, *The Conway polynomial in $\mathbb{R}^3$ and in thickened surfaces: A new determinant formulation*, Journal of Comb. theory B **61** (1994) 237–259.

[J1]   F. Jaeger, *Nowhere-zero flow problems*, in: Topics in Graph Theory 3, Academic Press, London (1988) 70–95.

[J2]   F. Jaeger, *Even subgraphs expansions for the flow polynomial of cubic plane maps*, Journal of Comb. theory B **52** (1991) 259–273.

[JV1]   V.F.R. Jones, *Hecke algebra representation of braid groups and link polynomials*, Annals Math. **126** (1987) 335–388.

[JV2]   V.F.R. Jones, *On knot invariants related to some statistical mechanical models*, Pacific J. Math **137** (1989) 311–334.

[KW]   M. Kac, J.C. Ward, *A Combinatorial solution of the two-dimensional ising model*, Physical Review **88** (1952) 1332–1337.

[KL]   M. Kang and M. Loebl, *The enumeration of planar graphs via Wick's theorem*, Advances in Mathematics (2009).

[KKLLL] N. Karmarkar, R. Karp, R. Lipton, L. Lovász, M. Luby, *A Monte-Carlo algorithm for estimating the permanent*, SIAM J. Comput. **22** (1993) 284–293.

[K]   C. Kassel, *Quantum Groups*, Springer-Verlag (1995).

[KF]   P.W. Kasteleyn and C.M. Fortuin, *Phase transitions in lattice systems with random local properties*, J.Phys.Soc.Japan **26** (Suppl.)(1969), 11–14.

[K1]   L.H. Kauffman, *State models and the Jones polynomial*, Topology **26** (1987).

[K2]   L.H. Kauffman, *Statistical mechanics and the Jones polynomial*, In: Braids, Contemp.Math.Pub. **78**, Am.Math. Soc. (1988) 263–297.

[KW]   H.A. Kramers, G.H. Wannier, Phys.Rev. **60** (1941) 252–262.

[KS]   L. Kauffman and P. Saleur, *Free fermions and the Alexander-Conway polynomial*, Comm. Math. Phys. **141** (1991) 293–327.

[KG]   G. Keller, *Equilibrium States in Ergodic Theory*, Cambridge University Press 1998.

[KSV]   A.Yu.Kitaev, A.Shen, M.N.Vyalyi, *Classical and quantum computation*, American Mathematical Society (2002).

[LW]   X-S. Lin and Z. Wang, *Random walk on knot diagrams, colored Jones polynomial and Ihara-Selberg zeta function*, Knots, braids, and mapping class groups–papers dedicated to Joan S. Birman (New York, 1998).

[LLW]   N. Linial, L. Lovász and A. Wigderson, *Rubber bands, convex embeddings and graph connectivity*, Combinatorica **8** (1988) 91–102.

[L1]   M. Loebl, *Chromatic polynomial, q-binomial counting and colored Jones function*, Advances in Mathematics **211** (2007).

[L2]   M. Loebl, *A discrete non-pfaffian aproach to the ising problem*, DIMACS, Series in Discrete Mathematics and Theoretical Computer Science, **63** (2004).

[LV]   M. Loebl and J. Vondrák, *Towards a theory of frustrated degeneracy*, Discrete Mathematics **271** (2003) 179–193.

[LL]   L. Lovász, *Discrete Analytic Functions: a survey*, in: Eigenvalues of Laplacians and other geometric operators (ed. A.Grigoriyan, S.T.Yau), International Press, Surveys in Differential Geometry IX (2004).

[LLV]   L. Lovász and K. Vesztergombi, *Geometric representations of graphs*, in:Paul Erdős and his Mathematics (ed. G.Halász, L. Lovász, M. Simonovits, V.T. Sós), Bolyai Soc.Math.Stud. **11**, János Bolyai Math.Soc. Budapest(2002) 471–498.

[LM]   M. Lothaire, *Combinatorics on words*, Encyclopedia of Mathematics and its applications, Addison-Wesley (1983).

[MD]   I.G. MacDonald, *Symmetric functions and Hall polynomials*, Oxford Math. Monographs, second edition (1995).

[MKS]   W. Magnus, A. Karrass, D. Solitar, *Combinatorial Group Theory*, Dover Publications (1976).

[MN]   J. Matoušek and J. Nešetřil, *Invitation to Discrete Mathematics*, Oxford University Press (1998).

[MJ]   J. Matoušek, *Using Borsuk-Ulam theorem*, Springer Verlag (2003).

[MIC]   M. McIntyre and G. Cairns, *A new formula for the winding number*, Geometriae Dedicata **46** (1993) 149–160.

[MC]   C. Mercat, *Discrete riemann surfaces and the ising model*, Commun. Math. Phys. **218** (2001) 177–216.

[MS]   S. Mertens, private communication.

[MT]   B. Mohar and C. Thomassen, *Graphs on surfaces*, The John Hopkins University Press (2001).

[NW]   S.D. Noble, D.J.A. Welsh, *A weighted graph polynomial from chromatic invariants of knots*, Annales de i'Institute Fourier **49** (1999) 101-131.

[O]   L. Onsager, Phys.Rev. **65** (1944) 117-149.

[P]   J. Propp, *The many faces of alternating-sign matrices*, Discrete Mathematics and Theoretical Computer Science (2001).

[RST]   N. Robertson, P.D. Seymour and R. Thomas, *Permanents, Pfaffian orientations and even directed circuits*, Annals of Mathematics **150** (1999) 929–975.

[RSST]   N. Robertson, D.P. Sanders, P.D. Seymour and R. Thomas, *The four-color theorem*, J. Combin. Theory Ser. B **70** (1997) 2–44.

[RP]   P. Rosenstiehl, *A New Proof of the Gauß Interlace Conjecture*, Advances in Applied Mathematics **23** (1999) 3–23.

[R]   W. Rudin, *Real and complex analysis*, McGraw-Hill. 3rd ed. (1987).

[SI]   I. Sarmiento, *The Polychromate and a chord diagram polynomial*, Annals of Combinatorics **4** (2000) 227–236.

[S]   P.D. Seymour, *Nowhere-zero 6 flows*, J. Combinatorial Theory(B) **30** (1981) 130–135.

[SAN]   A.N. Shiryayev, *Probability*, Springer-Verlag (1984).

[SA]   A.Sokal, *The multivariate Tutte polynomial (alias Potts model) for graphs and matroids*, preprint math.CO/0503607.

[S0]   R.P.Stanley, *Enumerative combinatorics I*, Cambridge University Press (1997).

[S1]   R.P. Stanley, *A Symmetric function generalization of the chromatic polynomial of a graph*, Advances in Mathematics **111** (1995) 166–194.

[S2]   R.P. Stanley, *Graph colorings and related symmetric functions: Ideas and applications*, Discrete Mathematics **193** (1998) 267–286.

[S3]   R.P. Stanley, *Enumerative Combinatorics volume I*, Cambridge University Press (1997).

[SM]   M. Sudan, *Algorithmic introduction to coding theory*, Lecture Notes (2001).

[TBD]   B.D.Thatte, *On the Nash-Williams'lemma in graph reconstruction theory*, Journal of Combinatorial Theory Ser.B **58** (1993) 280–290.

[TR]   R. Thomas, *An update on the four-color theorem*, Notices Amer.Math.Soc. **45** (1998) 848–859.

[TC]   C. Thomassen, *Whitney's 2-switching theorem, cycle space, and arc mappings of directed graphs*, J. Comb. Theory Ser. B **46** (1989) 257–291.

[TK]   K. Truemper, *On the delta-wye reduction for planar graphs*, Journal of Graph Theory **13** (1989) 141–148.

[TV]   V. Turaev, *The Yang-Baxter equation and invariants of links*, Inventiones Math. **92** (1988) 527–553.

[T1]   W.T. Tutte, *A ring in graph theory*, Proc. Camb. Phil. Soc. **43** (1947) 26–40.

[T2]   W.T. Tutte, *A contribution to the theory of chromatic polynomials*, Canad. J. Math. **6** (1954) 80–91.

[T3]   W.T.Tutte, *How to draw a graph*, Proc. London Math. Soc. **13** (1963) 743–767.

[WD]   D.J.A. Welsh, *Complexity: knots, colourings and counting*, London Math. Soc. Lecture Note Series **186** Cambridge Univ. Press (1993).

[WH1]   H. Whitney, *A logical expansion in mathematics*, Bull. Amer. Math. Soc. **38** (1932) 572–579.

[WH2]   H. Whitney, *On regular closed curves in the plane*, Compos. Math. **4** (1937) 276–284.

[WFY]   F.Y. Wu, *Knot theory and statistical mechanics*, Reviews of Modern Physics **64** (1992) 1099–1131.

[Z]   D. Zeilberger, *Proof of the alternating-sign matrix conjecture*, Electronic J. Comb. **3** (1996), R13.

List of Figures

Index

General and Algebraic Topology

Erich Ossa
Topologie

Eine anschauliche Einführung in die geometrischen und algebraischen Grundlagen

2., überarb. Aufl. 2009. X, 276 S. mit zahlr. Abbildungen. Ergänzendes Buchmaterial mit OnlinePlus. (Aufbaukurs Mathematik, hrsg. von Aigner, Martin / Gritzmann, Peter / Mehrmann, Volker / Wüstholz, Gisbert) Br. EUR 29,90
ISBN 978-3-8348-0874-5

Einführung - Allgemeine Topologie - Homotopie - Lie-Gruppen und homogene Räume - Homologie

Das Ziel des Buches ist eine umfassende Einführung sowohl in die geometrische wie die algebraische Topologie. Dabei werden lediglich gute Kenntnisse aus dem 1. Studienjahr in der Mathematik vorausgesetzt, die über die Analysis und lineare Algebra kaum hinausgehen; alle weiteren Hilfsmittel, wie die Grundbegriffe der mengentheoretischen Topologie, die Theorie der topologischen Gruppen und die algebraischen Grundlagen werden ebenfalls ausführlich dargestellt. Im Vordergrund stehen jedoch nicht die hieraus hervorgehenden technischen Apparate, sondern die geometrischen Fragestellungen, die erst den Anlass zu ihrer Entwicklung gaben.

VIEWEG+ TEUBNER

Abraham-Lincoln-Straße 46
65189 Wiesbaden
Fax 0611.7878-400
www.viewegteubner.de

Stand September 2009.
Änderungen vorbehalten.
Erhältlich im Buchhandel oder im Verlag.